aus der Reihe:

Innovationen mit Mikrowellen und Licht

Forschungsberichte aus dem Ferdinand-Braun-Institut, Leibniz-Institut für Höchstfrequenztechnik

Band 76

Nils Werner

Optical Feedback Aspects of Hybrid Integrated Diode Laser Systems

Herausgeber: Prof. Dr.-Ing Patrick Scheele

Ferdinand-Braun-Institut
Leibniz-Institut
für Höchstfrequenztechnik (FBH)
Gustav-Kirchhoff-Straße 4
12489 Berlin

Tel. +49.30.6392-2600
Fax +49.30.6392-2602

E-Mail fbh@fbh-berlin.de
Web www.fbh-berlin.de

Innovations with Microwaves and Light

Research Reports from the Ferdinand-Braun-Institut, Leibniz-Institut für Höchstfrequenztechnik

Preface of the Editor

Research-based ideas, developments, and concepts are the basis of scientific progress and competitiveness, expanding human knowledge and being expressed technologically as inventions. The resulting innovative products and services eventually find their way into public life.

Accordingly, the *"Research Reports from the Ferdinand-Braun-Institut, Leibniz-Institut für Höchstfrequenztechnik"* series compile the institute's latest research and developments. We would like to make our results broadly accessible and to stimulate further discussions, not least to enable as many of our developments as possible to enhance everyday life.

Hybrid integrated diode laser systems offer customized emission characteristics for a wide range of applications. However, each optical component can be a source of optical feedback that can adversely affect the system performance. In this thesis, the generation and the effects of optical feedback in hybrid integrated diode laser systems are modeled and investigated in experiments. With periodically poled nonlinear crystals, semiconductor optical amplifiers, and distributed Bragg-reflector ridge waveguide lasers, some of the most common components of hybrid integrated diode laser systems are studied systematically. The investigations provide a deeper understanding of optical feedback and approaches to its mitigation and control.

I wish you an informative and inspiring reading

Prof. Dr.-Ing. Patrick Scheele
Scientific Managing Director

The Ferdinand-Braun-Institut

The Ferdinand-Braun-Institut researches electronic and optical components, modules and systems based on compound semiconductors. These devices are key enablers that address the needs of today's society in fields like communications, energy, health and mobility. Specifically, FBH develops light sources from the visible to the ultra-violet spectral range: high-power diode lasers with excellent beam quality, UV light sources and hybrid laser systems. Applications range from medical technology, high-precision metrology and sensors to optical communications in space and integrated quantum technology. In the field of microwaves, FBH develops high-efficiency multi-functional power amplifiers and millimeter wave frontends targeting energy-efficient mobile communications as well as car safety systems. In addition, compact atmospheric microwave plasma sources that operate with economic low-voltage drivers are fabricated for use in a variety of applications, such as the treatment of skin diseases.

The FBH is a competence center for III-V compound semiconductors and has a strong international reputation. FBH competence covers the full range of capabilities, from design to fabrication to device characterization.

In close cooperation with industry, its research results lead to cutting-edge products. The institute also successfully turns innovative product ideas into spin-off companies. Thus, working in strategic partnerships with industry, FBH assures Germany's technological excellence in microwave and optoelectronic research.

Optical Feedback Aspects of Hybrid Integrated Diode Laser Systems

vorgelegt von
M. Sc.
Nils Werner
ORCID: 0000-0002-3055-6670

an der Fakultät IV - Elektrotechnik und Informatik
der Technischen Universität Berlin
zur Erlangung des akademischen Grades

Doktor der Naturwissenschaften
- Dr. rer. nat. -

genehmigte Dissertation

Promotionsausschuss:

Vorsitzender:	Prof. Dr. Wolfgang Heinrich
Gutachter:	Prof. Dr. Günther Tränkle
Gutachter:	Prof. Dr. Carlo Holly
Gutachter:	Prof. Achim Peters, Ph.D.

Tag der wissenschaftlichen Aussprache: 13. Oktober 2023

Berlin 2024

Bibliografische Information der Deutschen Nationalbibliothek

Die Deutsche Nationalbibliothek verzeichnet diese Publikation in der Deutschen Nationalbibliografie; detaillierte bibliographische Daten sind im Internet über http://dnb.d-nb.de abrufbar.

1. Aufl. - Göttingen: Cuvillier, 2024
Zugl.: Berlin, Technische Universität, Diss., 2024

Nonnenstieg 8, 37075 Göttingen
Telefon: 0551-54724-0
Telefax: 0551-54724-21
www.cuvillier.de

1. Auflage, 2024
Gedruckt auf umweltfreundlichem, säurefreiem Papier aus nachhaltiger Forstwirtschaft.
ISBN 978-3-68952-005-2
ISBN eBook 978-3-68952-040-3

Zusammenfassung

Hybridintegrierte Diodenlasersysteme sind Schlüsselkomponenten für Anwendungen in den verschiedensten Bereichen. Immer dann, wenn die Emissionseigenschaften einzelner Diodenlaser nicht ausreichen, werden sie in hybriden Systemen durch weitere optische Bauelemente ergänzt und so das Fähigkeitsspektrum erweitert. Beispiele sind optische Verstärker zur Leistungserhöhung oder Kristalle zur nichtlinearen Frequenzkonversion.

Die Kombination optischer Baugruppen führt jedoch zur gegenseitigen Beeinflussung, da jede Komponente eine Quelle optischen *Feedbacks* sein kann. Um die Komponenten optimal abzustimmen, ist es notwendig, optisches *Feedback* und dessen Auswirkungen genau zu erforschen. In dieser Arbeit werden die Untersuchungen experimentell und theoretisch an periodisch gepolten Kristallen (PPNC), halbleiterbasierten optischen Verstärkern (SOA) und Rippenwellenleiterlasern mit Bragg-Gitter (DBR-RWL) durchgeführt.

Die periodische Polung in PPNCs bedingt auch eine ungewollte regelmäßige Modulation des Brechungsindex. Es konnte gezeigt werden, dass das optische *Feedback* daher wie bei Bragg-Gittern wellenlängenabhängig ist und charakteristische Resonanzen mit Reflektivitäten um 10^{-2} auftreten. Abseits der Resonanzen wird das optische *Feedback* durch Toleranzen der periodischen Polung bestimmt, wobei sich Reflektivitäten um 10^{-4} ergeben. Das optische *Feedback* ist vorwiegend zur einstrahlenden Lichtquelle gerichtet.

Bei SOAs hängt das optische *Feedback* vom Arbeitspunkt ab. Schon bei geringen Ausgangsleistungen kann die *Feedback*-Leistung durch direkte Reflexion der Eingangsstrahlung und spontane Emission ein Verhältnis von 10^{-3} bis 10^{-2} relativ zur Eingangsleistung erreichen. Bei hohen Ausgangsleistungen steigt das Verhältnis auf über 10^{-1}. Gelangt externes optisches *Feedback* in den SOA, so erhöht sich vor allem das optische *Feedback* des SOAs, dessen Leistung die Eingangsleistung um ein Vielfaches übersteigen kann. Das räumliche Abstrahlverhalten des optischen *Feedbacks* wird durch die Struktur des SOAs bestimmt und ist typischerweise in Richtung der Eingangslaserquelle gerichtet.

Die Untersuchung von DBR-RWLs unter dem Einfluss externen optischen *Feedbacks* hat gezeigt, dass Effekte auch abseits üblicher *Feedback*-Betrachtungen auftreten. Abhängig von der Resonatorgeometrie und dem Arbeitspunkt des DBR-RWLs, treten neben Änderungen der Ausgangsleistung, Verschiebungen der Emissionswellenlänge oder *Coherence Collapse* auch spektrale Modensprünge und starke thermische Einflüsse auf. Ab externen optischen *Feedback*-Reflektivitäten um 10^{-3} können selbst bei DBR-RWLs, die eher unempfindlich gegenüber *Coherence Collapse* sind, störende Effekte auftreten.

Die Untersuchungen in dieser Arbeit ermöglichen ein tieferes Verständnis optischen *Feedbacks* und dessen Auswirkungen bei den einzelnen optischen Komponenten in hybrid integrierten Diodenlasersystemen. Dadurch können diese besser aufeinander abgestimmt werden. Aus den Erkenntnissen ergeben sich auch Ansätze für weitere Komponentenverbesserungen und neue Konzepte zur Realisierung hybrid integrierter Diodenlasersysteme.

Abstract

Hybrid integrated diode laser systems are key components for applications in a wide range of fields. Whenever the emission characteristics of individual diode lasers are not sufficient, diode lasers are supplemented by other optical components in hybrid systems to extend the range of capabilities. Examples are the combination with optical amplifiers for power enhancement or crystals for nonlinear frequency conversion.

However, the combination of optical components leads to mutual influencing, as each component can be a source of optical feedback. Therefore, in order to optimally match the components, it is essential to study both the feedback emission characteristics and the effects of optical feedback in detail. In this work, these studies are carried out experimentally and theoretically on periodically poled nonlinear crystals (PPNC), semiconductor optical amplifiers (SOA), and distributed Bragg-reflector ridge waveguide lasers (DBR-RWL).

The periodic poling in PPNCs also introduces an unwanted regular modulation of the refractive index. It has been shown that the optical feedback is therefore wavelength dependent, similar to Bragg-gratings, with characteristic resonances with reflectances around 10^{-2}. Away from the resonances, the reflectance is determined by the tolerances of the periodic poling, giving reflectances around 10^{-4}. The optical feedback is strongly directed toward the incident light source.

In SOAs, the optical feedback depends on the operating point. Even at low output powers, the optical feedback can be around a fraction of 10^{-3} to 10^{-2} of the input power, which can be caused by both direct reflection of the input radiation and spontaneous emission. At operating points with high output power, the optical feedback increases to a fraction greater than 10^{-1} of the input power. When external optical feedback enters the SOA, especially the feedback emission from the SOA increases and the corresponding power can exceed the input power many times over. The spatial emission characteristics of the optical feedback are essentially defined only by the internal structure of the SOA and are typically directed toward the input laser source.

Analysis of DBR-RWLs under the influence of strong external optical feedback has shown that effects beyond the usual feedback considerations also occur. Depending on the resonator geometry and the operating point of the laser, not only changes of the output power, shifts of the emission wavelength or coherence collapse occur, but also spectral mode hops and strong thermal effects. Starting from external feedback reflectances around 10^{-3}, even lasers insensitive to coherence collapse can be affected.

The investigations in this thesis provide a deeper understanding of the optical feedback characteristics of the individual components in hybrid diode laser systems and allow them to be better matched to each other. The results provide starting points for further component improvements and new concepts for the realization of hybrid laser systems.

Contents

1. Introduction

Hybrid integrated diode laser systems are key components for various demanding applications. In hybrid systems, diode lasers are combined with other optical components to realize tailored light sources, from single prototypes to commercial products.

By directly coupling of the emission of diode lasers into optical fibers [1], [2], the light can be made available via standardized interfaces, enabling integration into arbitrary applications. The pumping of fiber laser systems is only one example [3]. In hybrid master oscillator power amplifier (MOPA) systems a low power diode laser with specific emission characteristics is combined with a semiconductor optical amplifier (SOA). Thereby high-power laser sources with narrow spectral emission bandwidth [4]–[6] for atomic cooling and trapping applications [7] are realized. By exploiting nonlinear optics, the emission wavelength range of diode lasers can be extended. For example, while no suitable diode lasers are yet available in the yellow-green spectral range, corresponding emission wavelengths can be realized with systems consisting of near-infrared emitting diode lasers and periodically poled nonlinear crystals (PPNC) for second harmonic generation [8]–[10]. In particular, laser emission at 561 nm is required for high resolution microscopy [11], [12] in biomedical applications.

On the one hand, hybrid systems combine the respective advantages of different optical components and to achieve the required emission characteristics. On the other hand, each optical element can be a source of unwanted optical feedback. It has been shown early on in [13] that optical feedback has negative effects on the emission characteristics of diode lasers. In hybrid systems, both SOAs [14] and PPNCs [15] introduced optical feedback that negatively affected the emission characteristics of the laser.

Optical isolators are therefore used in hybrid systems to shield the respective laser source from optical feedback e.g. due to SOAs [6], [16], [17] or PPNCs [18]. However, there are several limitations in the use of optical isolators. While diode lasers allow a particularly high level of miniaturization of hybrid systems, miniaturized optical isolators are not available for all wavelengths. In addition, the available micro-optical isolators are limited to hundreds of milliwatts of maximum power, and transmission can be low depending on the wavelength range. Nevertheless, the isolation is finite so feedback effects can still be observed [19].

Optical feedback is an inherent aspect that has to be considered in the design of hybrid integrated diode laser systems. Therefore, it is necessary to analyze in detail the feedback emission characteristics of the optical components as well as the operating behavior under external feedback. PPNCs, SOAs, and diode lasers are studied in the most common assemblies of hybrid systems. In this way, the individual optical components can be optimally matched to each other in future developments. In addition, new approaches for optimizations arise in order to reduce optical feedback influences in hybrid integrated diode laser systems.

Thesis outline

This thesis is organized as follows: First, chapter 2 gives a general overview of hybrid integrated diode laser systems. In addition to the basics of the most commonly used components, examples of the occurrence of optical feedback effects are also shown. The subsequent detailed analysis of feedback aspects in hybrid integrated diode laser systems is divided into two parts.

The first part focuses on the sources of optical feedback. In chapter 3, the reflection properties of periodically poled nonlinear crystals (PPNC) for nonlinear frequency conversion are discussed. The optical feedback properties of PPNCs with different geometries and applications are investigated in various experiments and compared with a tailored theoretical description. Chapter 4 covers the optical feedback emitted by semiconductor optical amplifiers (SOAs). The feedback emission of two SOAs, a ridge waveguide amplifier (RWA) and a tapered amplifier (TPA), is characterized with respect to the operating conditions by experiments and numerical simulations. The results of the feedback studies of the first part are concluded in chapter 5 and discussed with respect to hybrid integrated diode laser systems.

The second part deals with the effects of optical feedback on typical diode laser sources. In chapter 6 the influences of external optical feedback on the operational behavior of TPAs are analyzed. The output emission characteristics as well as the feedback emission characteristics of the TPA are examined under different feedback conditions by means of experiments and numerical simulations. In chapter 7 the emission behavior of distributed Bragg-reflector ridge waveguide lasers (DBR-RWLs) with different resonator configurations exposed to optical feedback is studied. The experimental results are also compared with theoretical calculations. Chapter 8 concludes this second part and classifies the results in relation to hybrid diode laser systems.

Finally, chapter 9 gives the final conclusion of this work and points out future directions.

2. Hybrid Integrated Diode Laser Systems

In hybrid integrated diode laser systems, diode lasers are combined with other optical components to develop unique laser light sources tailored to the application requirements. The systems are characterized by a high potential for miniaturization and customization. Thereby, the spectral and spatial emission properties, such as the emission wavelength and the beam quality, can be designed almost arbitrarily. Furthermore, different operation modes are possible to achieve either continuous wave or pulsed emission. Thus, the compact and robust laser modules are promising alternatives to large and complex laser systems based on solid-state or dye lasers.

Figure 2.1 shows a photo of a hybrid integrated diode laser system in a module the size of a matchbox. The system is based on a diode laser that emits a single spatial and

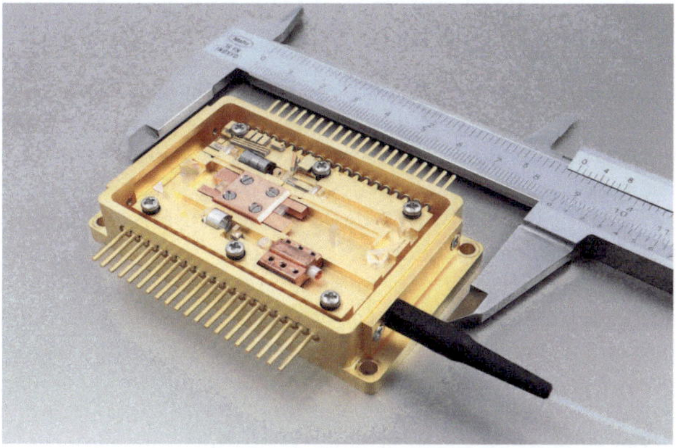

Figure 2.1.: Photo of a hybrid integrated diode laser system for watt-level output power in the yellow-green spectral region. The module features a hybrid master oscillator power amplifier diode laser source, second harmonic generation by a periodically poled nonlinear crystal, and an optical fiber output.

spectral mode at an emission wavelength of 1122 nm. Using a master oscillator power amplifier (MOPA) scheme, the laser emission is amplified by two orders of magnitude to several watt optical power in a semiconductor optical amplifier (SOA). In a periodically poled nonlinear crystal (PPNC), the near-infrared laser emission is efficiently converted to the visible yellow-green spectral region by nonlinear frequency conversion. At an emission wavelength of 561 nm, watt-level powers are realized. Finally, the visible laser light is coupled into an optical fiber. The light is shaped between the optical components by means of microlenses. The system shown is just one example that illustrates the combination of various active and passive optical components in a compact module.

In the next sections, the fundamentals of high-power diode laser sources and PPNCs, two commonly used components in hybrid integrated diode laser systems, are briefly presented. In addition, two examples of the occurrence of optical feedback effects in hybrid integrated diode laser systems are presented.

2.1. High-Power Diode Laser Sources

Diode lasers are typically based on epitaxially grown vertical layers of n- and p-doped semiconductors that form a p-n junction. By applying a forward bias voltage, a population inversion between the conduction band and the valence band is created in the active zone, i.e. the junction. Thereby, optical gain for the stimulated emission of radiation is provided. The focus in this work is on edge-emitting diode lasers based on gallium arsenide compound semiconductors. In these diode lasers, the vertical epitaxial layer structure also forms a planar waveguide for the radiation surrounding the active zone. The cleaved and optically coated facets can then be used as resonator mirrors. By selecting the material composition, emission wavelengths in the range of 620 nm to 1200 nm can be achieved [20]–[22].

Diode lasers have many advantages that clearly distinguish them from other types of lasers. Direct electrical pumping at low voltages allows simplified control of the laser operation, including direct modulation capability. In addition, their electro-optical efficiency is exceptionally high and their small size makes them ideal for micro-integration. However, the operating principle of diode lasers also brings challenges that must be considered in the development of diode laser systems.

To realize the emission of only one single spatial mode, the diode laser has to be structured accordingly. In addition to the vertical waveguide, the light must also be guided in the horizontal direction, for example by forming a ridge waveguide. However, in these ridge waveguide lasers (RWL), the optical output power is usually limited to around 1 W by damage thresholds due to high optical peak intensities [23], [24].

Due to the large number of possible band transitions between the conduction and valence bands, the optical gain extends over a spectral range of several nanometers. Therefore, spectrally selective structures such as a distributed Bragg-reflector (DBR) are monolithically integrated into the diode laser and act as a resonator mirror instead of the facet. When integrated into an RWL, the laser is referred to as DBR-RWL and single spectral mode emission with a spectral width around 1 MHz is achieved [5], [25].

To overcome the limited optical power of lasers emitting a single spatial and spectral mode, the master oscillator power amplifier (MOPA) concept can be used. The laser acts as the master oscillator while its emission is coupled into a semiconductor optical amplifier (SOA). This increases the optical power without significantly degrading the spectral and spatial beam characteristics. In addition, the spatial separation of the diode laser and the SOA provides the advantage of power scaling without changing the operating point of the diode laser. SOAs are generally similar to diode lasers except for the absence of a resonator. Depending on the required output power level, different types of SOAs are used. Ridge waveguide amplifiers (RWA) are similar to RWLs and provide watt-level output power. In tapered amplifiers (TPA) the light is not entirely guided in the horizontal direction allowing the light to diverge. Since the optical intensity is thereby reduced, much higher optical powers up to 10 W are possible [26], [27].

2.2. Frequency Conversion using Periodically Poled Nonlinear Crystals

Nonlinear frequency conversion is based on the fact that the electrical polarization in a medium is generally a nonlinear function of the local electric field. When excited by an oscillating electric field, such as a monochromatic electromagnetic wave, additional frequencies appear in the resulting oscillation of the polarization. Since an oscillating electric polarization is a source for the emission of electromagnetic radiation, waves with different frequencies can thus be generated from the incident wave.

The nonlinear dependence of the material polarization $\mathcal{P}$ on the electric field strength $\mathcal{E}$ is usually expressed as a power series in the electric field strength [28]

$$\mathcal{P}(t) = \varepsilon_0 \chi^{(1)} \mathcal{E}(t) + \varepsilon_0 \chi^{(2)} \mathcal{E}^2(t) + \varepsilon_0 \chi^{(3)} \mathcal{E}^3(t) + \mathcal{O}\left(\mathcal{E}^4(t)\right). \tag{2.1}$$

While ε_0 is the vacuum permittivity, the coefficients $\chi^{(i)}$ are the ith order electrical susceptibilities of the medium. In general, both the polarization and the electric fields are vectors, and the susceptibilities are tensors of rank $i+1$. Furthermore, the polarization may not respond to temporal variations of the electric field instantaneously, which makes the relation more complex. However, for most practical cases with electromagnetic waves, the simple scalar equation (2.1) is sufficient, noting that the susceptibilities depend on the orientation of the medium relative to the incident wave.

The exploitation of high second-order susceptibility $\chi^{(2)}$ in certain crystalline materials is widely used and enables a variety of different frequency conversion processes such as second harmonic generation (SHG), sum frequency generation (SFG), or spontaneous parametric down conversion (SPDC). However, efficient frequency conversion is generally limited by the dephasing of the electrical polarization and the radiated frequency-converted waves.

The effect is illustrated by the example of SHG, which generates light with a doubled frequency. Assuming a monochromatic wave propagating in z-direction and oscillating with frequency ω and wave vector k, the second-order term of the polarization oscillates with 2ω and $2k$. The corresponding radiated wave oscillates with 2ω but wave vector k_{SH}. Thus, after a distance Δz a phase difference [29] of

$$\Delta\phi = (k_{\mathrm{SH}} - 2k)\Delta z \tag{2.2}$$

is accumulated. Since the wave vectors $k = 2\pi n/\lambda$ and $k_{\mathrm{SH}} = 2\pi n_{\mathrm{SH}}/\lambda_{\mathrm{SH}}$ contain the refractive indices n and n_{SH} at the respective wavelengths λ and $\lambda_{\mathrm{SH}} = \lambda/2$, the phase difference is generally non-vanishing due to dispersion. When the phase difference reaches $\Delta\phi = \pi$, the generated waves superimpose destructively.

Minimizing the phase difference or at least the occurrence of destructive superpositions is called phase matching and is mandatory for efficient frequency conversion. The most common phase matching schemes are birefringent phase matching (BPM) and quasi phase matching (QPM). For ferroelectric crystals such as lithium niobate (LN), lithium tantalate (LT) or potassium titanyl phosphate (KTP), QPM is widely used. In QPM, the crystal structure is periodically inverted in the plane perpendicular to the propagation direction of the incident light beam (z), resulting in a sign change of the second-order susceptibility. Consequently, the second-order polarization term is phase shifted by π and generated second harmonic (SH) waves can still constructively superimpose. The period Λ_{QPM} of the periodic poling [29] is then given by

$$\Lambda_{\mathrm{QPM}} = \frac{\lambda}{2\left(n_{\mathrm{SH}} - n\right)}. \tag{2.3}$$

With the periodic poling the phase matching condition can be defined as

$$\Delta k = 2k - k_{\mathrm{SH}} - \frac{2\pi}{\Lambda_{\mathrm{QPM}}}, \tag{2.4}$$

where ideal phase matching consequently yields a phase mismatch $\Delta k = 0$. According to the dispersion and the temperature dependence of the refractive index, phase matching can only be guaranteed within a small wavelength and temperature range, typically far below 1 nm and 1 K, respectively [30].

Periodically poled nonlinear crystals (PPNC) are commercially available and offer several advantages for hybrid integrated systems. The QPM principle allows to exploit the highest second-order susceptibility and achieve sufficient conversion efficiency even in a simple single pass configuration. In contrast to other phase matching schemes, no spatial walk-off occurs.

Depending on the intended application and material properties, PPNCs with different geometries are used. Due to the quadratic dependence on the fundamental electric field, the conversion efficiency raises with increasing field strength, i.e. higher optical intensity of the incident beam. Thus, PPNCs with waveguide structures for strong optical confinement provide the highest conversion efficiencies [31]. However, the maximum achievable optical power is limited by parasitic effects due to the high optical intensity [32]. In contrast, bulk PPNCs are less sensitive to limiting effects, but require well-defined beam shaping of the incident beam, which is ideally a Gaussian beam [33].

2.3. Occurrence of Optical Feedback

Hybrid integrated diode laser systems can contain a variety of different active and passive optical elements. In addition to the optical elements with considerably varying reflective characteristics, reflections outside the actual laser system can also be a source of external optical feedback. Therefore, the effects of optical feedback in hybrid integrated diode laser systems can manifest themselves in a variety of ways.

In figure 2.2a the photo of a typical hybrid integrated master oscillator power amplifier (MOPA) laser source is shown. The MOPA is intended as a laser source with a narrow-

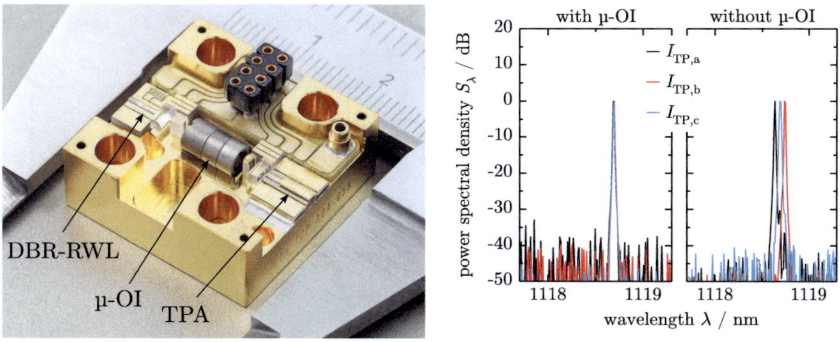

(a) (b)

Figure 2.2.: (a) Photo of a hybrid integrated master oscillator power amplifier (MOPA) comprising a distributed Bragg-reflector ridge waveguide laser (DBR-RWL) as the master oscillator, a micro-optical isolator (µ-OI), and a tapered amplifier (TPA). The MOPA is designed to provide multiple watts of optical output power in a single spectral mode. (b) Emission spectra of a hybrid integrated MOPA with and without µ-OI at three working points $I_{TP,a}$, $I_{TP,b}$ and $I_{TP,c}$ of the TPA. The DBR-RWL is operated at a constant working point.

band emission spectrum with a fixed central emission wavelength near 1120 nm and variable output power up to several watts. A distributed Bragg-reflector ridge waveguide laser (DBR-RWL) provides laser emission with the desired spectral characteristics which is shaped and coupled into a semiconductor optical amplifier (SOA) by microlenses. Here, the SOA is realized as a tapered amplifier (TPA). A micro-optical isolator (µ-OI) is placed between the DBR-RWL and the TPA. While the DBR-RWL is usually operated at a constant working point, the working point of the TPA is varied to scale the optical output power.

In figure 2.2b the measured emission spectra of a MOPA are given. The left plot shows the measured emission spectra at three working points of the TPA. The obtained optical output power increases from 2.0 W at the working point $I_{TP,a}$ to 2.3 W at the working point $I_{TP,c}$. The corresponding emission spectra remain unchanged as desired. However, when the µ-OI is removed, a different behavior is observed. The emission spectra of the same MOPA without µ-OI, operated at the same working points, are shown in the graph on the right. At each working point still a single spectral mode is observed but the central emission wavelengths differ by about 50 pm. With respect to the internal spectral mode spacing of the DBR-RWL laser cavity the emission wavelength changes can be associated with spectral mode hops. Hence, it can be assumed that the TPA is a source of optical feedback which, if not blocked by an optical isolator, causes spectral mode hops in the DBR-RWL.

Figure 2.3a displays a photo of hybrid integrated laser system with a periodically poled nonlinear crystal (PPNC) for second harmonic generation (SHG). The laser system

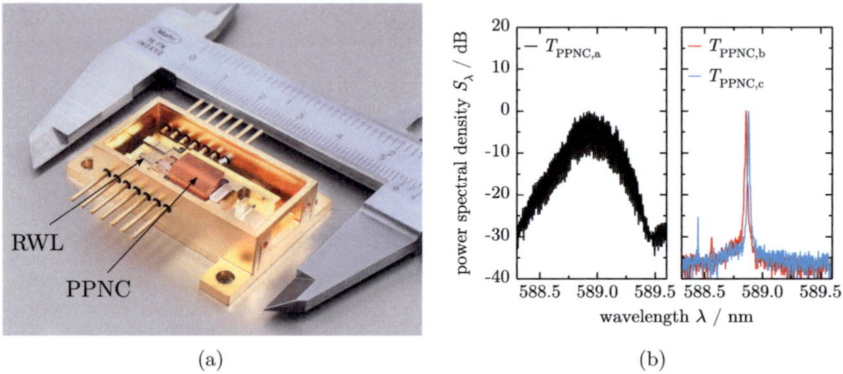

Figure 2.3.: (a) Photo of a hybrid integrated diode laser system comprising a ridge waveguide laser (RWL) and a periodically poled nonlinear crystal (PPNC) for second harmonic generation. The module is designed for spectral broadband emission in the yellow spectral region. (b) Emission spectra of the hybrid laser system at three different temperatures of the PPNC $T_{PPNC,a}$, $T_{PPNC,b}$ and $T_{PPNC,c}$. The RWL is operated at a constant working point.

is developed to provide spectrally broadband laser emission with a central emission wavelength of 589 nm. A ridge waveguide laser (RWL) is used as the pump laser source whose emission is directly coupled into a ridge waveguide PPNC. Due to the absence of spectrally selective elements in the Fabry-Pérot cavity of the RWL, the emission spectrum can be broad, only defined by the spectral distribution of the optical gain. Consequently, an emission spectrum is obtained that contains numerous emission lines extending over several nanometers around a central wavelength of 1178 nm. In the PPNC, the laser emission is converted to the yellow spectral region by SHG. In order to achieve ideal phase matching at 589 nm, the crystal temperature is adjusted accordingly.

In the left diagram in figure 2.3b the spectrum of the SH emission at the ideal working point with the crystal temperature $T_{PPNC,a}$ is given. Here, a spectrally broadband emission with a spectral width around 0.3 nm is realized. As the PPNC temperature changes, it is found that the emission spectrum changes drastically, although the RWL is still operated at the same working point. In the right diagram in figure 2.3b the emission spectra at two different crystal temperatures $T_{PPNC,b}$ and $T_{PPNC,c}$ are shown. The spectra are characterized by only one single emission line with a spectral width below

10 pm. By increasing the crystal temperature the central emission wavelength is slightly red-shifted from initially 588.85 nm to 588.87 nm. Therefore, it can be assumed that the PPNC generates spectrally selective optical feedback. The RWL is thus spectrally stabilized so that essentially only one longitudinal mode with a central wavelength of 1177.7 nm is emitted.

In the following chapters, PPNCs, SOAs, and DBR-RWLs, the typical components of hybrid integrated diode laser systems, are analyzed with respect to optical feedback. Thereby, the strength of the optical feedback of SOAs is determined and the feedback conditions under which spectral mode hops occur in DBR-RWLs. Furthermore, it is resolved why the optical feedback of PPNCs is spectrally selective.

Part I.

Sources of Optical Feedback

3. Optical Feedback of Periodically Poled Nonlinear Crystals

In this chapter, the optical feedback generated by periodically poled nonlinear crystals (PPNC) for nonlinear frequency conversion is discussed. While quasi phase matching (QPM) can be realized with various materials, the analysis focuses on ferroelectric crystals such as lithium niobate (LN) and lithium tantalate (LT), which are commonly used for single-pass frequency conversion in hybrid integrated laser sources. The theoretical and experimental investigations are presented for PPNCs optimized for second harmonic generation (SHG) as an example, but the results are also valid for PPNCs designed for other second-order nonlinear processes.

PPNCs are typically fabricated from a bulk crystal substrate that a priori has no reflective structure. The intention of QPM is to modulate only the second-order susceptibility [34], which is realized by periodically inverting the orientation of ferroelectric domains [35] in the crystal. The domain inversion can be realized by in-diffusion [36] or exchange [37] of ions or by application of external electric fields [38], [39]. The periodic structure is usually defined by lithographically applied masks or electrodes. Electric field poling has become the most common process, allowing periodic domain inversion of even millimeter-thick substrates, as required for bulk PPNCs, for example. However, the poling process not only modulates the second-order susceptibility, but also the linear susceptibility, i.e. the refractive index, can be affected. When periodic modulations of the refractive index occur, the structure acts like a Bragg-grating.

In the case of ion in-diffusion or ion exchange processes, the local material composition varies periodically due to the variation of ion concentrations. This results in a refractive index modulation. For PPNCs with channel waveguide characteristic reflectance spectra [40], [41] with a maximum reflectance as high as nearly unity [42] have been reported.

In the electric field poling process, the material decomposition of the crystal remains unchanged. Nevertheless, perturbations of the refractive index in the range of 10^{-5} to 10^{-3} have been found at the domain walls (DW), i.e. the interfaces between two opposing ferroelectric domains [43]–[47]. Here, the refractive index changes are related to several local effects at the DW, including strain [43], [48] and defects in the crystal structure [49]. The width of the DW region with refractive index perturbations is thinner than the typically several micrometers thick domains [50] and even of the order of several nanometers [47]. Therefore, PPNCs made by electric field poling can also act as a Bragg-grating. Typical diffraction at the periodic structure is observed for light incident parallel to the DW [45], [51]. In a SHG setup, where the pump light propagates perpendicular to the DW, the total reflectance of a bulk PPNC has been determined to the order of 10^{-4}[15]. However, the detailed spectral and spatial reflection characteristics of PPNCs have not been investigated so far.

The reflection characteristics of PPNCs depend on the crystal geometry and the beam shape of the pump light. The periodic refractive index variation in bulk PPNCs is similar to that of volume Bragg-gratings (VBG) [52], while the incident pump light is a focused Gaussian beam in order to achieve optimal conversion efficiency [33]. For VBGs, significant effects of the focusing condition of the incident Gaussian beam on the reflection properties have already been shown [53]. A comparable behavior can be expected for the reflection characteristics of bulk PPNCs.

The geometry and beam guiding of channel waveguide PPNCs is very comparable to that of fiber Bragg-gratings (FBG) [54]. Thus, typical FBG reflection characteristics [55] should be also obtained for channel waveguide PPNCs. In addition to the refractive index perturbations of the solitary periodic poling structure, subsequent processing steps for the waveguide formation can induce further index variations at the waveguide surface [56]. Similar to the effect of tilted gratings [57], this localized and possibly asymmetric refractive index perturbation can partially convert the guided mode into other waveguide modes [58]. Furthermore, the materials used for the waveguide cladding and the geometry of the waveguide also affect the reflection characteristics [59].

In the following sections, the spectral and spatial reflection characteristics of PPNCs with bulk and channel waveguide geometries and different design wavelengths in the near-infrared spectral region are investigated. Starting with theoretical considerations, the reflection properties of PPNCs are derived. The second section describes the selected PPNC samples, while the measurement details are covered in the third section. In the fourth section the experimental results are presented and compared with calculations based on the theoretical model. Finally, the results are summarized.

3.1. Theoretical modeling

In general, the calculation of the reflection properties of Bragg-gratings requires models that account for multiple reflections and depletion of the incident light, such as the coupled wave model [60], [61] or the transmission matrix approach [62]. However, due to the specific refractive index distribution of PPNCs, a simplified model can be used.

The refractive index in periodically poled nonlinear crystals (PPNC) can be assumed to be constant with perturbations only at the domain walls (DW) with a width much smaller than the domain width or light wavelength. Moreover, the index variation is the same for all DWs because the index profile is symmetric between inverted and non-inverted domains. Therefore, all DWs can be characterized by an effective amplitude reflectivity of the order of $r_{\mathrm{DW}} \approx 10^{-4}$[46], [47].

In figure 3.1 the schematic of a light beam with initial complex electric field amplitude E_{in} passing through a PPNC in z-direction is shown. Throughout the following considerations, the input facet of the crystal is assumed to be at $z = 0$. The length of the crystal is l and Λ_{QPM} is the period of the periodic poling. Due to the reflections at the numerous DWs, a feedback beam with electric field amplitude E_{FB} is generated. The reflection with field amplitude E_p originating from the pth DW is shown as an ex-

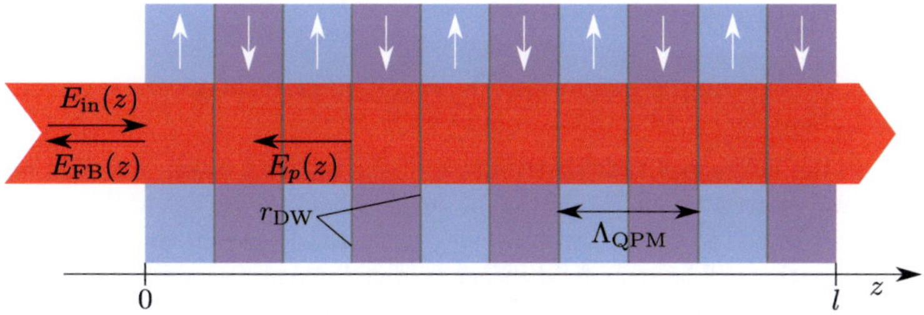

Figure 3.1.: Schematic of the reflection of a light beam at a periodically poled nonlinear crystal (PPNC). E_{in}: electric field amplitude of the incident light beam, E_{FB}: electric field amplitude of the feedback beam, E_p: electric field amplitude of the reflection at the pth domain wall (DW) r_{DW}: DW amplitude reflectivity, Λ_{QPM}: poling period, l: PPNC length.

ample. Since the DW are indistinguishable in terms of the refractive index profile, the period length of the resulting Bragg-grating is half the poling period. However, due to its importance for PPNCs, the poling period will always be used in the following.

3.1.1. Weakly Reflecting Bragg-Gratings

In the low reflectance limit [63], in [62] referred to as the Fourier limit, the electric field amplitude of the reflected feedback beam E_{FB} can be calculated under the condition of the following assumptions. First, only one reflection is considered at each DW while multiple reflections between the DWs are neglected. Second, the optical power of the incident light beam is constant throughout the crystal, which means that attenuation due to the reflections as well as nonlinear conversion processes is negligible. Third, the total reflectance of the PPNC is assumed to be small $R_{\text{PP}} \ll 1$. Then the feedback beam is the superposition of the individual reflections at the DWs. Consequently, the electric field amplitude of the feedback beam is defined by the sum [63] of the electric field amplitudes E_p of the reflections at the pth DW:

$$E_{\text{FB}} = \sum_{p=1}^{M} E_p. \tag{3.1}$$

M is the total number of DWs. Each contribution E_p is modeled as a plane reflection of the incident wave with field amplitude E_{in} at the DW with amplitude reflectivity r_{DW}

$$E_p(x, y, z) = r_{\text{DW}} E_{\text{in}}(x, y, -z + \Delta z_p). \tag{3.2}$$

The negative sign in the z-dependence of E_{in} accounts for the reversal of the propagation direction due to the reflection, and $\Delta z_p = p\Lambda_{\text{QPM}}$ is the geometric path difference of the reflection at the pth DW. Thus, the field amplitude of the feedback beam is generally defined by

$$E_{\mathrm{FB}}(x, y, z) = r_{\mathrm{DW}} \sum_{p=1}^{M} E_{\mathrm{in}}(x, y, -z + \Delta z_p). \tag{3.3}$$

Depending on the field amplitude distribution of the incident beam, the field amplitude distribution and spatial beam properties of the feedback beam can differ. The spatial beam properties of the feedback beam are characterized based on the near and far field intensity distributions. The near field intensity distribution S_{NF} describes the feedback beam profile in the region close to the PPNC and is defined as

$$S_{\mathrm{NF}}(x, y, z) = \frac{1}{2}\varepsilon_0 c_0 \left|E_{\mathrm{FB}}(x, y, z)\right|^2. \tag{3.4}$$

With the vacuum permittivity ε_0 and the vacuum speed of light c_0. The corresponding far field intensity distribution is obtained by the two-dimensional Fourier transform $\mathcal{F}$ of the field amplitude:

$$S_{\mathrm{FF}}(\theta_x, \theta_y) = \left|\mathcal{F}\{E_{\mathrm{FB}}\}(k_0 \sin(\theta_x), k_0 \sin(\theta_y))\right|^2. \tag{3.5}$$

θ_x and θ_y are the far field angles while $k_0 = 2\pi/\lambda$ is the vacuum wave vector with the wavelength λ of the incident beam. Integrating the intensity distributions of the feedback and incident light beams gives the total reflectance of the PPNC:

$$R_{\mathrm{PP}} = \frac{\iint |E_{\mathrm{FB}}(x, y)|^2 \mathrm{d}x\mathrm{d}y}{\iint |E_{\mathrm{in}}(x, y)|^2 \mathrm{d}x\mathrm{d}y}. \tag{3.6}$$

In the following, the reflected field amplitude and the corresponding total reflectance of PPNCs are calculated for plane waves first. The results are then used to derive the reflection properties for focused Gaussian beams in bulk PPNCs or guided modes in channel waveguide PPNCs.

The complex electric field amplitude of a plane wave traveling in the z-direction inside the crystal is given by

$$E_{\mathrm{in,\,plane}}(z) = E_0 e^{-\imath kz}. \tag{3.7}$$

E_0 is the constant amplitude. The wave vector $k = nk_0$ of the incident beam contains the refractive index n of the crystal.

Inserting (3.7) into (3.3) leads to

$$E_{\mathrm{FB,\,plane}}(z) = e^{\imath kz} \sum_{p=1}^{M} r_{\mathrm{DW}} E_0 e^{-\imath kp\Lambda_{\mathrm{QPM}}}. \tag{3.8}$$

The sum in equation (3.8) is a finite geometric series that can be calculated directly:

$$E_{\mathrm{FB,\,plane}}(z) = r_{\mathrm{DW}} E_0 e^{\imath kz} \frac{e^{-\imath Mk\Lambda_{\mathrm{QPM}}} - 1}{1 - e^{\imath k\Lambda_{\mathrm{QPM}}}}. \tag{3.9}$$

Thus, the expression for the plane wave reflectance of the crystal reads

$$R_{\text{PP, plane}} = |r_{\text{DW}}|^2 \frac{\cos(Mk\Lambda_{\text{QPM}}) - 1}{\cos(k\Lambda_{\text{QPM}}) - 1}. \tag{3.10}$$

If the Bragg-condition is fulfilled and the arguments of the cosines become an integer multiple of 2π, both the numerator and the denominator will be zero. The limit of the fraction in (3.10) is still defined, giving

$$\max(R_{\text{PP, plane}}) = |r_{\text{DW}} M|^2, \tag{3.11}$$

as expected when all partial reflections resonantly add. The spectral positions λ_{r} of the resonances are defined by the Bragg-equation

$$\lambda_{\text{r}} = \frac{n\Lambda_{\text{QPM}}}{m}, \tag{3.12}$$

where m is the Bragg-order. Compared to the literature, equation (3.12) lacks a factor of two since the poling period is twice the Bragg-period of the grating. In section A.1 the correspondence of the used theoretical model with models for Bragg-gratings in the literature is given.

Due to the dependence on the wave vector k and the poling period Λ_{QPM} in equation (3.10), both the refractive index and the wavelength affect the reflectance equally. Therefore, adjusting the temperature of the crystal, which is a common procedure to achieve the exact phase matching condition, has a similar effect as changing the wavelength.

For a given poling period, the wavelengths at which resonances occur depend on the respective refractive indices. In contrast, the phase matching condition depends on the refractive index of all wavelengths involved, including that of the frequency-converted light (see equation (2.3)). Therefore, the position of the resonances does not correlate with the fundamental wavelength of the intended nonlinear process.

Reflection of Gaussian Beams

In order to obtain the reflection properties of a Gaussian beam at the PPNC, it is helpful to perform the calculation in the transverse Fourier space (k_x, k_y), similar to [64]. The transverse Fourier transform of the electric field distribution of a Gaussian beam has the form

$$E_{\text{in, Gauss}}(k_x, k_y, z) = E_0 \frac{w_0^2}{2} e^{-\frac{k_x^2+k_y^2}{4} w_0^2} e^{-\imath \left[k - \frac{k_x^2+k_y^2}{2k}\right](z - z_{\text{BW}})} \tag{3.13}$$

where w_0 and z_{BW} are the beam waist radius and the z-coordinate of the beam waist, respectively. The term in square brackets is the paraxial approximation of the z-component of the wave vector and is denoted in the following as $\tilde{k}_z = k - \frac{k_x^2+k_y^2}{2k}$. Since the z-dependence of the equation (3.13) is similar to that of the plane wave in (3.7), the plane wave results can be adapted directly. The field amplitude distribution of the reflected beam is then

$$E_{\mathrm{FB,\ Gauss}}(k_x, k_y, z) = r_{\mathrm{DW}} E_0 \frac{w_0^2}{2} e^{-\frac{k_x^2+k_y^2}{4} w_0^2} e^{\imath \tilde{k}_z (z+z_{\mathrm{BW}})} \frac{e^{-\imath M \tilde{k}_z \Lambda_{\mathrm{QPM}}} - 1}{1 - e^{\imath \tilde{k}_z \Lambda_{\mathrm{QPM}}}}. \tag{3.14}$$

To calculate the real space electric field distribution, inverse Fourier transforms have to be applied first by numerical methods.

Due to the Plancherel theorem the necessary integrals for the calculation of the reflectance in equation (3.6) can be performed equally in Fourier space. Using the plane wave result (3.10) again gives the final result for the reflectance:

$$R_{\mathrm{PP,\ Gauss}} = |r_{\mathrm{DW}}|^2 \frac{w_0^2}{2\pi} \iint e^{-\frac{k_x^2+k_y^2}{2} w_0^2} \frac{\cos(M \tilde{k}_z \Lambda_{\mathrm{QPM}}) - 1}{\cos(\tilde{k}_z \Lambda_{\mathrm{QPM}}) - 1} \mathrm{d}k_x \mathrm{d}k_y. \tag{3.15}$$

While the integral for the incident beam can be solved analytically, the integration for the reflected beam in equation (3.15) has to be solved numerically. It is important to note that the reflectance $R_{\mathrm{PP,\ Gauss}}$ does not depend on the position of the beam waist z_{BW} but only on the beam waist radius w_0. For large beam waist radii, the exponential term in equation (3.15) is non-vanishing only for $k_x = k_y \approx 0$ and the result converges to the plane wave result with $\tilde{k}_z \to k$.

Reflection of Guided Modes

The guided light propagating through a waveguide structure can be described by the superposition of orthogonal waveguide modes. The modes of channel waveguides are typically characterized by two indices, representing two independent degrees of freedom according to the geometry of the waveguide. For the sake of clarity, however, the modes will be denoted only by a single index ν. The electric field distribution of each mode is given by

$$E_\nu(x, y, z) = E_0 \Phi_\nu(x, y) e^{-\imath \beta_\nu z}. \tag{3.16}$$

The real valued function Φ_ν is the transverse mode profile, which is normalized as $\iint \Phi_\nu^2(x, y) \mathrm{d}x \mathrm{d}y = 1$. The propagation constant β_ν is defined by $\beta_\nu = 2\pi(n + \Delta n_{\mathrm{eff},\nu})/\lambda$ with the refractive index n of the waveguide core (in most cases the bulk PPNC material) and the effective index difference $\Delta n_{\mathrm{eff},\nu}$ for the respective mode ν.

With respect to optimal nonlinear conversion efficiency, the incident light is coupled so that only the fundamental mode $\nu = 0$ is excited and therefore $E_{\mathrm{in,\ modes}} = E_{\nu=0}$. In contrast, the reflected beam is a superposition of multiple modes due to possible conversion to higher-order modes according to asymmetric refractive index variations of the DWs. The reflected field amplitude is thus given by

$$E_{\mathrm{FB,\ modes}}(x, y, z) = E_0 \sum_\nu r_{\mathrm{DW},\nu} \Phi_\nu(x, y) e^{\imath \beta_\nu z} \frac{e^{-\imath M \tilde{\beta}_\nu \Lambda_{\mathrm{QPM}}} - 1}{1 - e^{\imath \tilde{\beta}_\nu \Lambda_{\mathrm{QPM}}}}. \tag{3.17}$$

The DW amplitude reflectivity $r_{\mathrm{DW},\nu}$ describes the conversion from the fundamental mode to the respective mode ν. With the mode conversion the phase terms include the effective propagation constant $\tilde{\beta}_\nu = \frac{1}{2}(\beta_0 + \beta_\nu)$, which accounts for the fundamental forward and respective backward propagating mode ν.

The total reflectance of the PPNC is finally

$$R_{\text{PP, modes}} = \sum_\nu |r_{\text{DW},\nu}|^2 \frac{\cos(M\tilde{\beta}_\nu \Lambda_{\text{QPM}}) - 1}{\cos(\tilde{\beta}_\nu \Lambda_{\text{QPM}}) - 1} \tag{3.18}$$

where the normalization and orthogonality properties have been applied. Since the propagation constant depends not only on the refractive index of the crystal but also on the surrounding waveguide materials and geometry, the spectral position of the resonances may vary.

Exemplary Reflectance Spectra

To illustrate the differences in the spectral reflection characteristics for different incident beams and PPNC geometries, the respective reflectance as a function of the wavelength is shown in figure 3.2. An exemplary PPNC with a length of 10 mm and a poling period of 8 µm is assumed. In the case of lithium niobate, such a crystal is phase matched for second harmonic generation (SHG) at a fundamental wavelength around 1130 nm. The reflectance spectra are calculated by equations (3.10), (3.15) and (3.18) for three different incident beams: a plane wave, a Gaussian beam with 17 µm waist radius, and a guided mode with two reflected modes. For simplicity, the reference refractive index is assumed to be dispersionless with a constant value of 2.14. The effective index of the guided modes is set to the reference index for the fundamental mode, while the higher-order mode has an effective index reduced by a difference of 10^{-2}. The amplitude reflectivity of the DW is set to 10^{-4}, which is equally assumed for the guided modes. In figure 3.2a the reflectance spectra of the three incident beams are shown in linear scale. All spectra have a reflectance peak around 1141.2 nm corresponding to the 15th order Bragg-reflex. In addition, a second reflectance peak appears for the reflected higher-order mode at a wavelength red-shifted by around 2.5 nm. The maximum reflectance reaches nearly 6×10^{-2} in the case of plane waves or guided modes, whereas the maximum reflectance for the Gaussian beam is about half.

In figure 3.2b the structure of the reflectance spectra is resolved in more detail in logarithmic scale for two wavelength ranges. Around 1130 nm, where optimal phase matching for SHG is achieved, the reflectance of all beams drops to 10^{-8}. While the reflectance spectra for the plane wave and the guided modes are subject to strong oscillations, the reflectance spectrum of the Gaussian beam is not. With respect to the representation of the Gaussian beam in Fourier space in equation (3.13), the z-component of the wave vector varies with the transverse wave vectors k_x and k_y. Therefore, constructive and destructive interference is not possible for all transverse wave vectors simultaneously and the maxima and minima are less pronounced. The structure of the resonance maxima around 1141.2 nm differs significantly for the three beams. The plane wave peak is symmetric, whereas the maximum for the guided mode shows slight asymmetry and less pronounced oscillations due to the nearby second resonance maximum of the higher-order

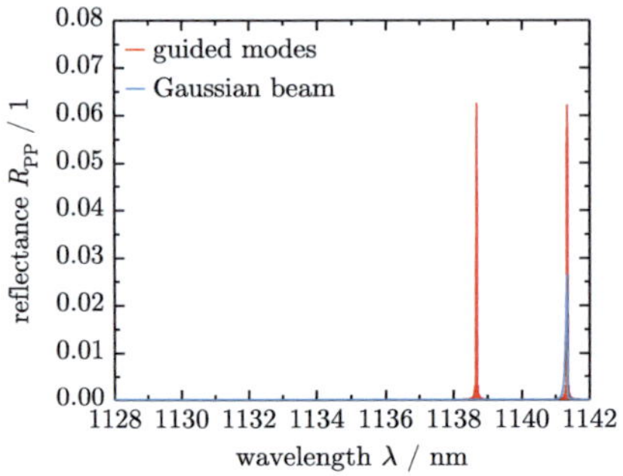

(a)

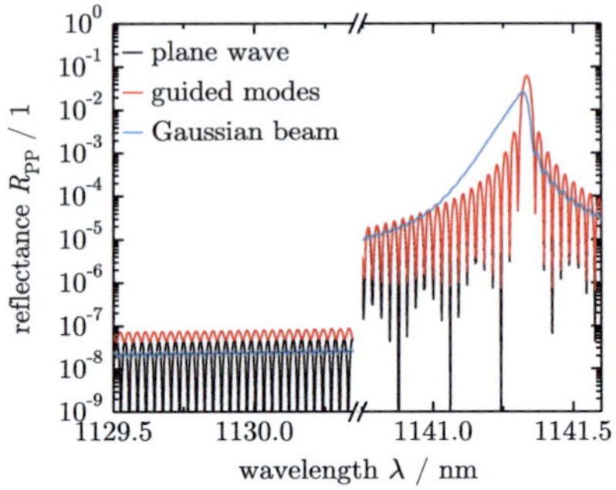

(b)

Figure 3.2.: Reflectance spectra of a 10 mm long periodically poled nonlinear crystal with a poling period of 8 µm for different incident beam shapes. (a) Reflectance spectra for a guided mode with two reflected modes and a Gaussian beam. (b) Reflectance spectra compared to the plane wave result in logarithmic scale at two exemplary wavelength regions in detail.

mode. The maximum of the reflectance spectrum of the Gaussian beam is broadened and strongly asymmetric with less distinct oscillations. The shape of the resonance maximum is consistent with the calculations for strongly reflecting VBGs reported in [53].

Influence of Fabrication Tolerances

Tolerances in the crystal fabrication process lead to various fluctuations in the periodic structure, with small variations in the DW positions being of particular interest. Random variations in the DW position induce random phase perturbations in the individual reflections. The reflectance spectra of Bragg-gratings with random phase noise can be significantly affected. Here, the reflectance offsite the resonance maxima is increased by several orders of magnitude while the maximum reflectance can be reduced [63], [65]–[67].

The perturbations of the DW positions can be divided into two types. First, the position of each DW varies independently around the ideal position, which is referred to as the duty cycle error. Second, the DW position varies with respect to the directly preceding DW position. Here, the variation in the position of each DW depends on all the preceding domains, which is similar to a random walk process. This type of domain error is called period error. The actual tolerances occurring for PPNCs can be assumed to be a combination of both error types. However, with respect to the actual manufacturing process, the occurrence of the error types may differ and only one error type may be dominant. For example, if the PPNC is manufactured by a holographic process, the long-range periodicity is precisely controlled and the duty cycle error will dominate.

In the following, the reflectance of PPNCs with perturbed poling is calculated for a combination of both error types. Again, the description is given first in the low reflectance approximation for plane waves. The calculation is comparable to [30] where the influence of poling perturbations on the conversion efficiency is derived, and a similar description for period errors can be found in [63]. The mathematical scheme given there can be used with minor adaptions.

In order to treat the influence of the stochastic variations of the DW positions, the ensemble average of the PPNC reflectance is calculated. According to equations (3.9) and (3.6) the average reflectance is given by

$$\langle R_{\text{PP, plane}} \rangle = |r_{\text{DW}}|^2 \left\langle \left| \sum_{p=1}^{M} e^{-\imath k \Delta z_p} \right|^2 \right\rangle . \tag{3.19}$$

The geometric path difference of the pth reflection is now defined as

$$\Delta z_p = p\Lambda_{\text{QPM}} + 2\delta_{\text{DE},p} + 2\sum_{q=1}^{p} \delta_{\text{PE},q} \tag{3.20}$$

where $\delta_{\text{DE},p}$ and $\delta_{\text{PE},p}$ are the duty cycle and period error deviations of the pth DW with respect to the ideal position. The variations of the DW position are assumed to be statistically independent each with a normal probability distribution with zero mean and standard deviations σ_{DE} and σ_{PE}.

By inserting the definition in (3.20) into (3.19), the ensemble average can be calculated analytically. While the full derivation is given in section A.2, the final result reads:

$$\begin{aligned} \langle R_{\text{PP, plane}} \rangle &= |r_{\text{DW}}|^2 \\ &\times \left(M + 2e^{-4k^2\sigma_{\text{DE}}^2} \Re \left\{ \frac{M-1}{e^{2k^2\sigma_{\text{PE}}^2 - \imath k \Lambda_{\text{QPM}}} - 1} + \frac{e^{-(M-1)(2k^2\sigma_{\text{PE}}^2 - \imath k \Lambda_{\text{QPM}})} - 1}{(e^{2k^2\sigma_{\text{PE}}^2 - \imath k \Lambda_{\text{QPM}}} - 1)^2} \right\} \right) . \end{aligned} \tag{3.21}$$

$\Re\{\cdot\}$ denotes the real part. As the standard deviations σ_{DE} and σ_{PE} approach zero, equation (3.21) becomes equation (3.10), the result for an unperturbed crystal. On the other hand, if σ_{DE} and σ_{PE} are large, equation (3.21) becomes $|r_{\text{DW}}|^2 M$ which is the reflectance if all reflections add incoherently.

The derived equation (3.21) can be adapted directly to incident Gaussian beams or guided modes. The wave vector k has to be changed to the respective value and the integration in Fourier space has to be performed:

$$\begin{aligned} \langle R_{\text{PP, Gauss}} \rangle &= |r_{\text{DW}}|^2 \frac{w_0^2}{2\pi} \iint e^{-\frac{k_x^2 + k_y^2}{2} w_0^2} \\ &\times \left(M + 2e^{-4k^2\sigma_{\text{DE}}^2} \Re \left\{ \frac{M-1}{e^{2\tilde{k}_z^2\sigma_{\text{PE}}^2 - \imath \tilde{k}_z \Lambda_{\text{QPM}}} - 1} + \frac{e^{-(M-1)(2\tilde{k}_z^2\sigma_{\text{PE}}^2 - \imath \tilde{k}_z \Lambda_{\text{QPM}})} - 1}{(e^{2\tilde{k}_z^2\sigma_{\text{PE}}^2 - \imath \tilde{k}_z \Lambda_{\text{QPM}}} - 1)^2} \right\} \right) \mathrm{d}k_x \mathrm{d}k_y . \end{aligned} \tag{3.22}$$

Similarly, the equation for guided modes reads

$$\langle R_{\text{PP, modes}} \rangle = \sum_{\nu} |r_{\text{DW},\nu}|^2 \times \left(M + 2e^{-4k^2\sigma_{\text{DE}}^2} \Re \left\{ \frac{M-1}{e^{2\tilde{\beta}_\nu^2\sigma_{\text{PE}}^2 - \imath\tilde{\beta}_\nu\Lambda_{\text{QPM}}} - 1} + \frac{e^{-(M-1)(2\tilde{\beta}_\nu^2\sigma_{\text{PE}}^2 - \imath\tilde{\beta}_\nu\Lambda_{\text{QPM}})} - 1}{(e^{2\tilde{\beta}_\nu^2\sigma_{\text{PE}}^2 - \imath\tilde{\beta}_\nu\Lambda_{\text{QPM}}} - 1)^2} \right\} \right). \tag{3.23}$$

It is important to note that the phase difference of the reflections at the individual DWs is also affected by varying propagation constants due to waveguide inhomogeneity. Since the phase shift of a reflection at a particular DW depends on all previous phase shifts in the optical path, the mathematical description is equivalent to the consideration of period errors. Thus, tolerances in the waveguide geometry result in similar distorted reflectance spectra as would be expected for poling tolerances with period error.

The effects of poling errors are exemplarily analyzed for plane waves and the same crystal geometry used before for the spectra in figure 3.2. A standard deviation of 4 nm is assumed for either the duty cycle error or the period error, while the other error type is set to zero. With such small DW position errors of only 0.1 % of the DW width ($\Lambda_{\text{QPM}}/2 = 4\,\mu\text{m}$) the nonlinear conversion efficiency is not noticeably affected. In section A.2, the relation for the conversion efficiency under the influence both error types is given in equation (A.26). The calculated conversion efficiency is lower by 10^{-3} % and 1 % for the duty cycle error and the period error type, respectively.

In figure 3.3 the calculated reflectance spectra are shown for the duty cycle error and period error types. The reflectance spectra for the crystal with ideal poling ($\sigma_{\text{DE}} = \sigma_{\text{PE}} = 0\,\text{nm}$) are also given as reference. In addition to the average reflectance defined by equation (3.21), the reflectance spectra are also calculated for a random sample of DW positions. Here, the sum in equation (3.8) is calculated numerically while using the path difference in equation (3.20) with random numbers for either $\delta_{\text{DE},pp}$ or $\delta_{\text{PE},pp}$.

Although the DW position errors are small enough not to affect the nonlinear conversion efficiency, the reflectance spectra are significantly changed. Figure 3.3a shows the reflectance spectra for pure duty cycle error. The average reflectance offsite the resonance maximum is increased by at least one order of magnitude. While the average reflectance is only characterized by tiny oscillations, the reflectance spectrum based on the random sample still shows strong variations. In contrast, the reflectance around the resonance maximum remains almost unchanged. Only the maximum reflectance is slightly reduced by a difference of 10^{-3}.

The reflectance spectra for DW positions with pure period error are given in figure 3.3b. Around 1130 nm, the reflectance spectra are comparable to the spectra for pure duty cycle error. The average reflectance is increased by a factor of ten and the reflectance oscillates strongly in the case of the random sample. At the resonance maximum, the reflectance spectra differ significantly. The average maximum reflectance directly at the resonance is reduced by one order of magnitude, whereas the reflectance near the resonance is increased. The reflectance spectrum based on the random sample shows multiple reflectance fluctuations. Here, several reflectance peaks with a maximum

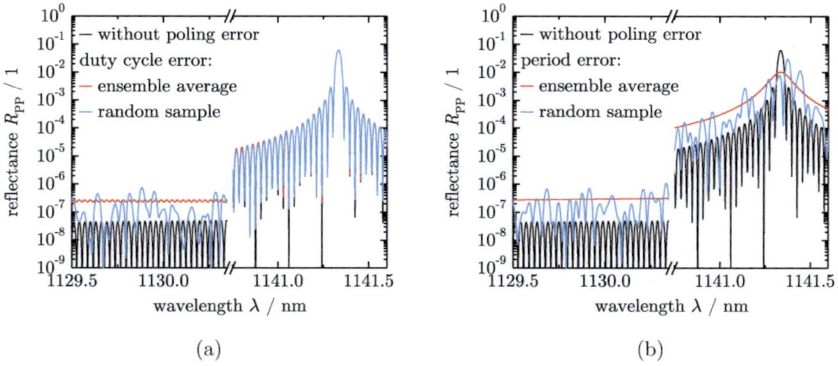

Figure 3.3.: Reflectance spectra of a 10 mm long periodically poled nonlinear crystal with a poling period of 8 µm and different periodic poling error types for plane waves. The reflectance spectra for perturbed poling are shown for the ensemble average and for a random sample. (a) Reflectance spectra for pure duty cycle error with a standard deviation of $\sigma_{DE} = 4$ nm. (b) Reflectance spectra for pure period error with a standard deviation of $\sigma_{PE} = 4$ nm.

reflectance in excess of 1×10^{-2} are obtained. Furthermore, the global maximum reflectance is not at the resonance wavelength. Although the maximum average reflectance is strongly reduced, the actual reflectance maximum can still be much larger.

3.2. Investigated Periodically Poled Nonlinear Crystal Samples

The periodically poled nonlinear crystal (PPNC) samples used to investigate the generated optical feedback are selected in order to cover a variety of different effects. With the characterization of a bulk PPNC the intrinsic reflective properties of the periodic poling are analyzed. The results serve as a reference for the investigation of PPNCs with channel waveguide structures and allow to distinguish influences due to the presence of the waveguides. In addition to different waveguide types, different PPNC materials are also investigated.

3.2.1. Bulk Periodically Poled Nonlinear Crystals

The optical feedback generated by bulk PPNCs is characterized in terms of the spectral reflection characteristics as well as the spatial intensity distribution of the optical feedback. A commercially available PPNC made of congruent lithium niobate (LN) doped with 5 mol % magnesium oxide by *HC Photonics Corp.* is used for the experiments. The PPNC is designed for second harmonic generation (SHG) at a fundamental wavelength of 1122 nm.

In figure 3.4 the schematic of the bulk PPNC is shown. The bulk PPNC has a length

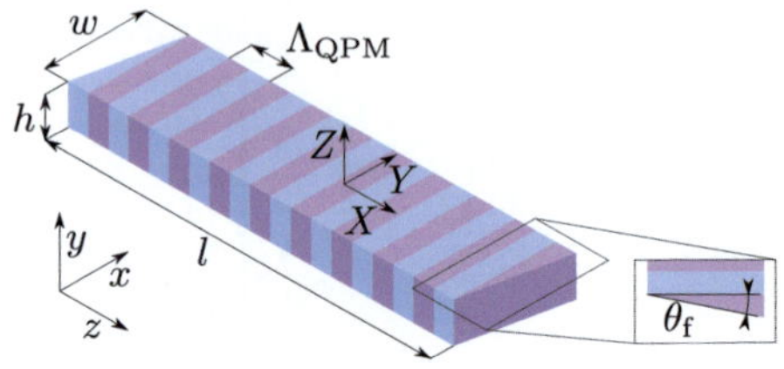

Figure 3.4.: Schematic of a bulk periodically poled nonlinear crystal (PPNC). l: PPNC length, w: PPNC width, h: PPNC height, Λ_{QPM}: poling period, θ_f: facet angle, X, Y, Z: crystallographic axes.

$l = 10\,\text{mm}$ while the width w and height h are 2 mm and 0.5 mm, respectively. The poling period is specified as $\Lambda_{QPM} = 8.19\,\mu\text{m}$ which is optimized for phase matching at a fundamental wavelength of 1122 nm at a temperature of 30 °C. To avoid parasitic reflections from the bulk PPNC end surfaces, the facets are polished under an angle of $\theta_f = 6°$ and coated with an anti-reflective coating with a specified residual reflectivity of less than 0.3 %.

The crystal polar-axis Z is aligned along the y-axis of the laboratory coordinate system defined in figure 3.4 and the crystallographic X-axis is collinear with z. This orientation of the PPNC is also referred to as z-cut. Thus, the input beam has to be polarized along the y-direction in extraordinary polarization in order to achieve SHG. According to the length of the PPNC, the ideal waist diameter of the incident Gaussian beam is calculated [33] to be about 34 µm.

A similar bulk PPNC has been used in [68] to generate 340 mW of second harmonic (SH) laser emission from more than 6 W of nearly diffraction limited pump power. The corresponding conversion efficiency is determined to be $1.6\,\%\,\text{W}^{-1}$. The reflectivity of the domain walls (DW) for a comparable bulk PPNC type has been investigated in [47], yielding a value of $-66\,\text{dB}$ and hence an amplitude reflectivity around 5×10^{-4}.

By use of equation (3.12), involving the given poling period and the refractive index defined by the Sellmeier equation (A.27) with parameters for LN from table B.1, resonances around 1097 nm and 1171 nm can be expected for 16th and 15th order Bragg-reflections. According to equation (3.11), a maximum reflectance at the resonances of the order of 10^{-2} can be estimated from the DW amplitude reflectivity and the number of DWs.

3.2.2. Channel Waveguide Periodically Poled Nonlinear Crystals

The optical feedback generated by channel waveguide PPNCs is investigated using three different commercially available PPNCs. The PPNCs differ in poling period, waveguide geometry and material composition.

Two PPNCs manufactured by *HC Photonics Corp.* are made of congruent LN doped with magnesium oxide and feature a ridge waveguide (RW). The RW PPNCs differ only in their poling periods, which are optimized for SHG at 1122 nm and 1178 nm, respectively. The third PPNC from *Oxide Corp.* is made of magnesium oxide-doped stoichiometric lithium tantalate (LT) with a poling period optimized for SHG at 1064 nm. Here, the waveguide is a circular depressed cladding waveguide (DCW).

A schematic of the RW and DCW PPNCs is shown in figure 3.5. The RW PPNC

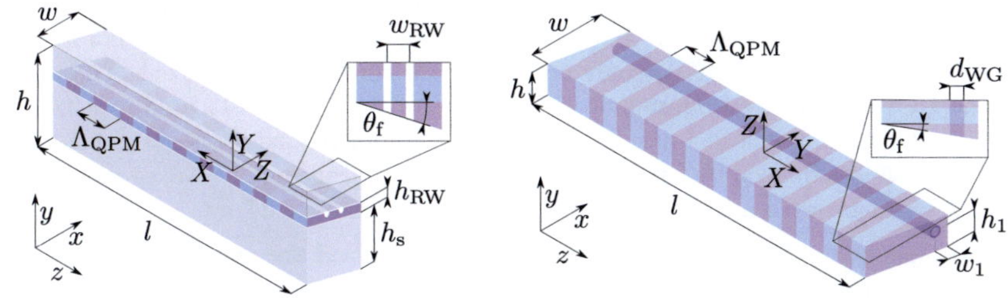

Figure 3.5.: Schematic of channel waveguide periodically poled nonlinear crystals (PPNC) with (a) ridge waveguide (RW) and (b) depressed cladding waveguide (DCW). l: PPNC length, w: PPNC width, h: PPNC height, Λ_{QPM}: poling period, w_{RW}: RW width, θ_f: facet angle, h_{RW}: RW planar layer height, h_s: substrate thickness, d_{WG}: DCW diameter, h_1: distance from DCW center to bottom edge, w_1: distance from DCW center to side edge, X, Y, Z: crystallographic axes.

is shown in figure 3.5a. The length of the PPNC is $l = 10\,\text{mm}$ and the width w and height h are 0.5 mm and 1.5 mm, respectively. Parasitic reflections from the PPNC facets towards the laser source are avoided by angled polishing of the PPNC end faces with an angle of $\theta_f = 6°$ in combination with an antireflection coating with a residual reflectivity of less than 0.5 %. In the vertical direction (y), the light is guided in a thin layer with a thickness of $h_{RW} = 3.7\,\mu\text{m}$, which is covered by a silicon oxide cladding. The horizontal (x-direction) guiding is realized by a RW which is defined by two trenches at a distance of a $w_{RW} = 9.5\,\mu\text{m}$. While the RW is centered horizontally in the PPNC, the waveguide is embedded asymmetrically vertically between two substrates. The lower substrate has a thickness of $h_s = 1\,\text{mm}$. The resulting mode field diameter for the fundamental mode is specified to be 3 µm by 6 µm for the vertical and horizontal directions, respectively.

The periodic poling is realized in the thin waveguide layer only with poling periods Λ_{QPM} of 7.72 µm and 8.82 µm. With these poling periods, the phase matching condition is reached for 1122 nm at 25 °C and for 1178 nm at 36 °C. The PPNC orientation is y-cut with its polar-axis Z aligned with the x axis of the laboratory coordinate system and Y parallel to the y direction. Consequently, the input beam has to be polarized in the x-direction.

In [69] similar PPNCs have been investigated. At 1122 nm more than 500 mW SH light generated with 1.2 W pump power have been shown. The conversion efficiency is $230\,\%\,W^{-1}$. An even higher output power of 860 mW has been realized with the PPNC with a design wavelength of 1178 nm. Taking into account the pump power of 2.1 W, a slightly reduced conversion efficiency of $190\,\%\,W^{-1}$ is obtained when compared to the PPNC for 1122 nm.

In figure 3.5b the schematic of the DCW PPNC is shown. The PPNC has a length of $l = 30$ mm, a width of $w = 2$ mm and a height of $h = 0.5$ mm. The PPNC facets are polished at an angle of $\theta_f = 2°$ and coated to achieve a reflectivity below 0.5 %. While the PPNC has multiple DCWs, only one of them is further characterized. The position of the DCW is shown in figure 3.5b. The center of the DCW is located at a height of $h_1 = 0.36$ mm and at a distance of $w_1 = 0.12$ mm from the side edge. The cladding of the waveguide is defined by femtosecond-laser writing, resulting in a waveguide diameter of $d_{WG} = 36$ µm with a corresponding mode field diameter of 24 µm.

Throughout the DCW PPNC, periodic poling is realized and oriented as z-cut. Thus, the crystal axes Z (polar-axis) and Y are aligned with the laboratory axes y and x, and the input beam has to be polarized in the y-direction. With a poling period of $\Lambda_{QPM} = 7.98$ µm SHG can be achieved at a fundamental wavelength of 1064 nm and a temperature of 36 °C.

The conversion efficiency is specified to be $16.7\,\%\,W^{-1}$. Comparable PPNCs with slightly different dimensions are presented in [70], [71]. With the large waveguide diameter, pump power levels well above 1 W are possible without the risk of damage.

To estimate the reflection characteristics of the channel waveguide PPNCs, information about the guided modes are required. In section A.5, the calculated effective refractive index differences and far field intensity distributions of the guided modes for both channel waveguide geometries are given. According to the polarization of the incident light and the geometry of the RW or the DCW, the guided light is described by the TE and LP modes, respectively. The effective index difference of the fundamental TE_{00} mode for the RW PPNCs is around -6×10^{-3}. Due to the high refractive index contrast between the core and cladding regions and the respective waveguide dimensions, higher-order modes are also guided. With the DCW PPNC, multiple guided modes are also possible, with the LP_{01} fundamental mode having an effective index difference around -1×10^{-4}.

With the effective index differences, the Sellmeier equation (A.27) and the material parameters given in tables B.1 and B.2, the resonance wavelengths can be estimated by equation (3.12). For the PPNC designed for 1122 nm conversion, resonance can be expected around 1101 nm, corresponding to the 15th order Bragg-reflection. Thus, the resonances occur at wavelengths several nanometers away from the design wavelength, similar to the bulk PPNC presented in section 3.2.1. In contrast, the expected resonances for the PPNCs with design wavelengths of 1178 nm and 1064 nm are in the region of the design wavelength. Here the resonances at 1177 nm (16th order) and 1064 nm (16th order) can be determined. Assuming the same DW amplitude reflectivity reported for the bulk PPNC, the maximum reflectance of the channel waveguide PPNCs is similarly estimated by equation (3.11) to be of the order of 10^{-2}.

3.3. Experimental Setup and Methods

The periodically poled nonlinear crystals (PPNC) are characterized under the same operating conditions that occur in a typical setup for second harmonic generation (SHG). In figure 3.6 the schematic of the experimental setup is shown. Three different laser sources

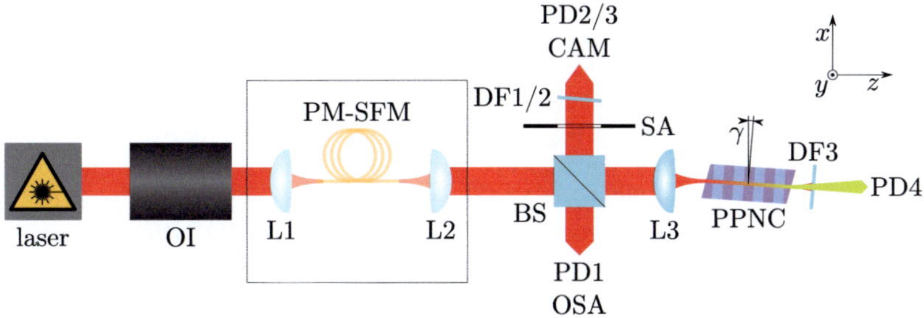

Figure 3.6.: Schematic of the experimental setup for the characterization of periodically poled nonlinear crystals (PPNC). OI: optical isolator, L1 / L2: optional aspherical lenses, PM-SMF: optional polarization maintaining single mode fiber, BS: beamsplitter, L3: aspherical or spherical lens, SA: slit aperture, DF1 - DF3: dichroic filters, PD1 - PD4: power detectors, CAM: camera, γ: PPNC tilt angle.

with adjustable emission wavelengths are selected according to the design wavelength of the PPNC. Either an external-cavity diode laser (ECDL) or distributed Bragg-reflector ridge waveguide lasers (DBR-RWL) are used. Additional information on the laser sources is given in section C.1. The collimated beam from each laser source passes through an optical isolator (OI) that provides an isolation of at least 40 dB. In the case of the bulk and DCW PPNC, an additional combination of lenses L1 and L2 with focal lengths of 8 mm and 15.3 mm and a 2 m long polarization maintaining single mode fiber (PM-SMF) with a mode field diameter of 6.6 µm is used as a mode filter. The beamsplitter (BS) reflects a part of the incident light to a power detector PD1 (*XLP12-3S-H2*) or an optical spectrum analyzer (OSA). The reflectance of the beamsplitter R_{BS} is polarization and wavelength dependent and is therefore calibrated individually for each measurement. While the reflectance is assumed to be constant within the tuning range of each laser source, reflectances in the range of 0.325 to 0.70 are obtained. The transmitted light is coupled into the PPNC which is realized by the lens L3. A spherical lens with a focal length of 60 mm and a numerical aperture of 0.2 is used for the bulk and DCW PPNCs, whereas an aspherical lens with a focal length of 8 mm and a numerical aperture of 0.5 couples the light into the RW PPNCs. In the case of the bulk PPNC, the beam waist position is set to the center of the PPNC. The PPNCs are mounted on a heat sink that is actively temperature controlled. In order to compensate for diffraction at the angled facets, the PPNCs are rotated in the $x-z$ plane by a certain tilt angle γ.

The PPNC feedback is reflected from the BS and passes through a dichroic filter (DF1) and a slit aperture (SA), which are used to separate the different the contributions of the feedback beam. Feedback from the bulk and RW PPNC facets is reflected at an angle of 25.9° relative to the incident light. Similarly, the angle for the DCW PPNC is 8.6°. According to the numerical aperture of the lens L3, the facet reflection does not pass through the lens in the case of the bulk PPNC. This also applies to residual reflections at the output facet of the bulk PPNC, even if multiple total internal reflections occur at the side facets of the PPNC. In contrast, the facet reflection of the channel waveguide PPNCs passes through the lens L3, resulting in a collimated beam with a distinct offset from the optical axis. The reflections due to the periodic poling propagate backwards on-axis. Thus, the slit aperture can be used to block each reflection individually. With the long pass filter DF1, possible emissions of the second harmonic (SH) light are blocked. The transmission is $T_{\mathrm{DF1}} = 0.889$, while the visible light is attenuated by 40 dB. The power of the feedback is measured with the power detector PD2 (*PH20-Ge*), and the near field and corresponding far field intensity distributions are recorded with a camera (CAM). The intensity distributions of the bulk and DCW PPNCs are recorded with the *Grasshopper3 GS3-U3-91S6M* camera while the distributions of the other PPNCs are obtained with the *Scorpion SCOR-20-SOM* camera. The lenses used for the imaging onto the camera are not shown in figure 3.6, but the measurement setup is described in section C.4. The near field intensity distribution is obtained by imaging the center of the bulk PPNC or the input facet in the case of the channel waveguide PPNCs.

The generated SH light is characterized in terms of optical power behind the PPNC. The dichroic filter DF3 attenuates the fundamental pump light by 60 dB, while $T_{\mathrm{DF3}} = 0.947$ of the SH light is transmitted and measured by the power detector PD4 (*PH100-Si*). In addition to the feedback at the fundamental wavelength, the feedback emission of the generated SH light is investigated. Here, the dichroic filter DF1 is replaced by a short pass filter DF2 with the same specifications as DF3, whereas the power detector PD3 (*PH100-Si*) is used instead of PD2. Since the SH light is generated inside the PPNC, no reflection at the front facet occurs and the slit aperture is not used.

The operating conditions for the PPNCs are given in table 3.1. The horizontal (x) and vertical (y) beam waist diameters of the incident light are calculated from the spatial emission characteristics of the laser sources or the PM-SMF and the focal lengths of the coupling lenses L3. With respect to the different laser light sources, the optical power in front of the PPNC is at most about 50 mW. At such low input power levels, neither depletion of the input power due to SHG nor thermal effects are significant.

For each PPNC, the setup is aligned first using the optical power P_{SH} of the generated SH light as the optimization criterion. Thereby ideal coupling of the incident light into the PPNC is ensured. The power of the generated SH light is calculated by $P_{\mathrm{SH}} = P_{\mathrm{PD4}}/T_{\mathrm{DF3}}$ with the measured power P_{PD4} of power detector PD4 and the transmission T_{DF3} of the dichroic filter DF3. The reflectance spectra of the PPNCs are measured by determining the reflectance at distinct wavelengths and PPNC temperatures. The wavelength tuning capability of the 1064 nm laser source without spectral mode hops and the use of the PM-SMF as a mode filter allows the wavelength to be continuously varied without affecting the coupling into the DCW PPNC. Using automatic data acquisition,

Table 3.1.: Operating conditions used for the characterization of periodically poled nonlinear crystals (PPNC) with different geometries and design wavelengths. RW: ridge waveguide, DCW: depressed cladding waveguide, ECDL: external-cavity diode laser, DBR-RWL: distributed Bragg-reflector ridge waveguide laser.

PPNC geometry	bulk	RW	RW	DCW
design wavelength	1122 nm	1122 nm	1178 nm	1064 nm
laser source type	ECDL	ECDL	DBR-RWL	DBR-RWL
maximum input power $P_{\text{in, max}}$	51 mW	49 mW	46 mW	40 mW
horizontal beam waist diameter	26 µm	5.4 µm	10.5 µm	26 µm
vertical beam waist diameter	26 µm	3.1 µm	3.5 µm	26 µm
PPNC tilt angle γ	6.9°	6.9°	6.9°	2.3°
PPNC default temperature T_0	25 °C	25 °C	20 °C	40 °C

the reflectance can be measured with a much larger number of wavelength samples than is possible with the other two laser sources. The three contributions to the feedback (facet, periodic poling and SH light) are always distinguished and measured separately.

The PPNC reflectance R_{PP} of the periodic poling or reflectivity R_{f} of the back facet are calculated from the ratio of the total input power P_{in} to the feedback power P_{FB}. Using the measured power P_{PD1} from power detector PD1 and accounting for the reflectivity R_{BS} of the beamsplitter BS, the input power is given by $P_{\text{in}} = (1 - R_{\text{BS}})P_{\text{PD1}}/R_{\text{BS}}$. With the measured power P_{PD2} from power detector PD2 and the transmission T_{DF1} of DF1, the power of the feedback reads $P_{\text{FB}} = P_{\text{PD2}}/(R_{\text{BS}}T_{\text{DF1}})$. Consequently, the reflectance or reflectivity is obtained by

$$R_{\text{PP/f}} = \frac{1}{(1 - R_{\text{BS}})T_{\text{DF1}}} \frac{P_{\text{PD2}}}{P_{\text{PD1}}}. \tag{3.24}$$

In contrast, the reflectance R_{SH} of the PPNC for the SH light is defined as the ratio of the power $P_{\text{FB,SH}}$ of the reflected SH light to the power P_{SH} of the generated SH light. The power of the reflected SH light is calculated by $P_{\text{FB,SH}} = P_{\text{PD3}}/(R_{\text{BS}}T_{\text{DF2}})$. Here, P_{PD3} is the measured optical power with power detector PD3, while T_{DF2} is the transmission of the dichroic filter DF2, which is identical to that of DF3. The relation for the reflectance R_{SH} then reads

$$R_{\text{SH}} = \frac{1}{R_{\text{BS}}} \frac{P_{\text{PD3}}}{P_{\text{PD4}}}. \tag{3.25}$$

The power of the SH light inside the PPNC is not constant along the propagation direction and depends on the actual phase matching condition. Thus, the corresponding determined reflectance R_{SH} may not be comparable to the reflectance R_{PP} of the fundamental near infrared light.

With respect to the emission wavelength tuning characteristics of the individual laser sources, only a coarse and non-uniform wavelength sampling is possible in some cases. In contrast, the temperature of the PPNC can be precisely tuned with a step size of 0.1 K. By exploiting the temperature scaling properties, the reflectance spectra of the PPNCs can be determined with a much higher resolution and the accessible wavelength range can be extended beyond the wavelength range of the laser source. Near a resonance, a temperature change ΔT can be related to a wavelength change $\Delta\lambda$ by

$$\Delta\lambda = \partial_T \lambda_\mathrm{r} \Delta T, \tag{3.26}$$

where λ_r is the resonance wavelength (see equation (3.12)). Here, the temperature dependence of the refractive index and the thermal expansion, which changes the poling period, must be taken into account.

The temperature and wavelength dependence of the refractive index is modeled by Sellmeier equation, while thermal expansion is considered only in the longitudinal direction. The relations are given in the sections A.3 and A.4. The material parameters used for lithium niobate and lithium tantalate are provided in tables B.1 and B.2.

3.4. Results

The optical feedback emission of periodically poled nonlinear crystals (PPNC) is investigated with respect to its spectral and spatial properties. First, the results for a bulk PPNC are presented, followed by the results for channel waveguide PPNCs.

3.4.1. Optical Feedback of Bulk Periodically Poled Nonlinear Crystals

The spectral reflection characteristics of the bulk PPNC are characterized first. The measured reflectance of the PPNC as a function of the incident light wavelength are given in figure 3.7a. In addition to the measured reflectance, the spectral region where significant second harmonic generation (SHG) occurs is shown in light gray for comparison. The gray dashed line represents the normalized SHG conversion efficiency, which reaches unity at 1121.5 nm. The measured reflectance has a maximum of 7.0×10^{-3} in a narrow spectral region around 1098.2 nm. The inset of figure 3.7a shows the reflectance maximum in more detail, revealing a slight asymmetry. Except for the resonance maximum, the reflectance fluctuates around a value of about 10^{-3}. The single resonance maximum with the relatively high surrounding background reflectance indicates that the tolerances of the periodic poling are dominated by a duty cycle error.

In figure 3.7b the same measured data is compared with the calculation using equation (3.22) for poling tolerances with pure duty cycle error. With the domain wall (DW) amplitude reflectivity of $r_\mathrm{DW} = 6.7 \times 10^{-4}$ and the standard deviation of the DW position variation $\sigma_\mathrm{DE} = 92\,\mathrm{nm}$ obtained from a fit to the experimental data, the calculated reflectance spectrum agrees with the measured results. Both the shape of the resonance maximum and the reflectance off resonance (around 10^{-3}) agree with the model.

Compared to the dimensions of the periodic poling, the standard deviation of the DW position error is small at only 2.2 % of the DW width. With respect to SHG, the fitted standard deviation is too small to significantly affect the conversion efficiency. Therefore,

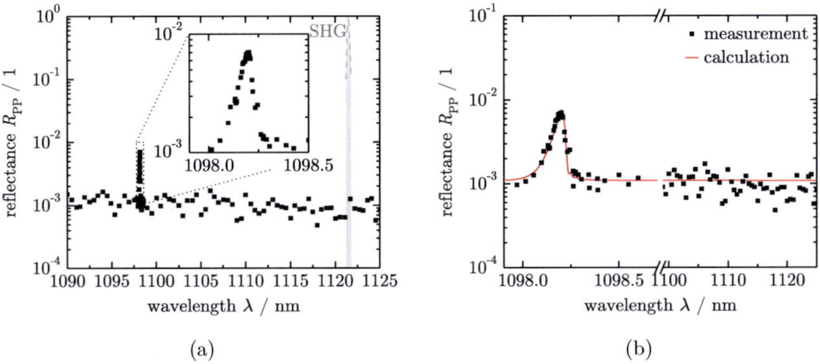

Figure 3.7.: (a) Measured reflectance of a bulk periodically poled nonlinear crystal as a function of the incident light wavelength. The dashed gray line symbolizes the normalized conversion efficiency for second harmonic generation (SHG). (b) The same measured reflectance compared to the calculated average reflectance including a pure duty cycle error.

the poling quality of the crystal can still be considered as sufficient with a calculated conversion efficiency reaching the ideal value of about 99.9 %.

To investigate the spatial emission characteristics of the crystal feedback, the intensity distributions of the near field at the center of the crystal and the corresponding far field are considered. At three exemplary wavelengths, the measured intensity distributions are compared with the calculations based on equation (3.14). The calculated near and far field intensity distributions are obtained from the calculated electric field amplitudes of the feedback beam in the absence of poling tolerances. The results are shown in figure 3.8. Both wavelengths, 1098.12 nm and 1098.21 nm, are in the spectral region of the resonance with the maximum reflectance obtained at the latter wavelength. In contrast, the wavelength 1121.55 nm is in the spectral region where a high SHG conversion efficiency is achieved.

The near field intensity distributions presented in figure 3.8a depend strongly on the wavelength of the incident light. While a Gaussian shaped spot with a diameter of about 50 µm and a weak surrounding circular diffraction fringe is observed at 1098.21 nm, the intensity distributions at the other two wavelengths are broader. Here, a central intensity maximum and several diffraction fringes are obtained. At 1098.12 nm the central maximum has a diameter of about 14 µm. Compared to the beam waist of the incident beam with a diameter of 26 µm, the observed diameter of the central lobe of the feedback varies by a factor of two. The calculations yield comparable intensity patterns at the corresponding wavelengths. Especially in the region of the resonance maximum a good agreement between measurement and calculation is achieved. At a wavelength of 1121.55 nm the calculated intensity pattern is still comparable to the measured re-

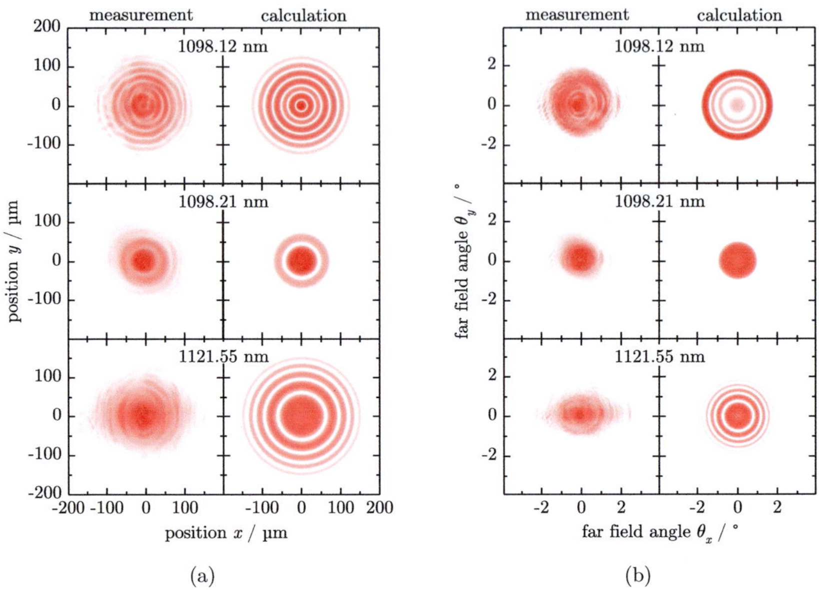

Figure 3.8.: Measured and calculated spatial intensity distributions of the reflection at a bulk periodically poled nonlinear crystal (PPNC) at three different wavelengths. (a) Near field intensity distributions at the center of the bulk PPNC. (b) Far field intensity distributions.

sult, but interference fringes with a large diameter above 200 µm are not visible in the measured intensity distribution.

In figure 3.8b the far field intensity distributions are shown. The actual intensity distribution depends again strongly on the wavelength. At 1098.21 nm there is a narrow far field pattern with a full width below 2°. Except for a local minimum in the center, no other minima are visible. In contrast, several interference fringes are observed at 1098.12 nm and 1121.55 nm. Despite the different far field intensity patterns, the maximum far field angles obtained are within the range of the incident beam which has a full far field angle of 3.1°. As with the near fields, the calculated intensity distributions are in good agreement with the measured data. While the agreement between measurement and calculation is best at 1098.21 nm, there are differences at the other two wavelengths. Although the overall width and general shape of the intensity distributions are comparable, the relative intensity of distinct interference fringes or the position of minima or maxima varies. Since the calculation of the intensity patterns is based on equation (3.14) and does not take poling tolerances into account, the differences between measurement and calculation are reasonable at least for 1121.55 nm. Here the reflection is strongly affected by the poling tolerances, in contrast to the reflections at wavelengths close to the resonance maximum.

3.4.2. Optical Feedback of Channel Waveguide Periodically Poled Nonlinear Crystals

The spectral and spatial reflection characteristics of all three channel waveguide PPNCs are investigated, beginning with the ridge waveguide (RW) PPNCs designed for 1122 nm and 1178 nm and ending with the depressed cladding waveguide (DCW) PPNC with a design wavelength of 1064 nm. Each PPNC is examined with a focus on specific effects.

The RW PPNC with a design wavelength of 1122 nm is characterized with respect to different feedback contributions and thermal effects. In figure 3.9a the reflectance is shown as a function of the wavelength. The insets display the far field intensity

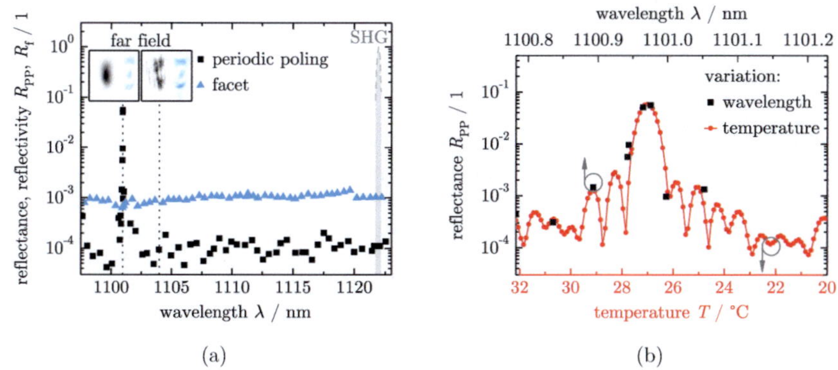

Figure 3.9.: (a) Measured periodic poling reflectance R_{PP} and facet reflectivity R_f of a ridge waveguide periodically poled nonlinear crystal (RW PPNC) with a design wavelength of 1122 nm as a function of incident light wavelength. The dashed gray line depicts the normalized second harmonic generation (SHG) conversion efficiency. The insets show the far field intensity distribution of the reflection at two wavelengths indicating different contributions in different color scales. (b) Reflectance spectrum of the RW PPNC obtained by varying the wavelength of the incident light at a fixed temperature of 25 °C and varying the temperature at a fixed wavelength of 1101.07 nm.

distribution of the RW PPNC feedback at two exemplary wavelengths of 1100.98 nm and 1103.99 nm, whereas the gray dashed line shows the normalized conversion efficiency, marking the wavelength region where significant SHG occurs.

From the far field intensity distributions, the reflections from the periodic poling and the facet can be clearly distinguished. The feedback due to the periodic poling is visualized by the gray color scale while the facet reflection is colored in blue. The intensity distribution of the facet reflection is characterized by interference fringes that are independent of wavelength. In contrast, the far field intensity distribution of the periodic poling reflection varies strongly with wavelength. Therefore, it can be assumed that the periodic poling feedback is a superposition of multiple waveguide modes. Although the intensity pattern changes with the wavelength, the reflection from the front facet is sufficiently separated and the two patterns do not overlap. Thus, it is possible to characterize each reflection individually.

Over the entire wavelength range, the reflectivity of the facet remains nearly constant around a value of 10^{-3}, which is consistent with the specification of the antireflection coating. The reflectance due to the periodic poling shows a more pronounced wavelength dependence. While the mean reflectance due to the periodic poling is an order of magnitude smaller than the facet reflectivity, a sharp resonance is obtained with a peak reflectance of 5.5×10^{-2} at a wavelength of 1100.98 nm. According to the far field intensity distribution at this wavelength, the reflection corresponds to the fundamental mode TE_{00} of the waveguide. From the resonance wavelength, the effective index difference of the mode can be determined to be $\Delta n_{\mathrm{eff}} = -6.2 \times 10^{-3}$. Both the resonance wavelength and the effective index difference agree with the theoretical considerations.

The shape of the resonance maximum of the periodic poling reflectance is further analyzed. In addition to varying the wavelength at a constant PPNC temperature, the dependence of the reflectance on the PPNC temperature at a fixed incident wavelength is investigated. The results of both measurements are given in figure 3.9b. The upper and lower abscissas are aligned with respect to the resonance wavelength, while the scale divisions are obtained from the temperature scaling (3.26) with $0.0366\,\mathrm{nm\,K^{-1}}$ at a reference wavelength and temperature of 1101 nm and 25 °C, respectively. Compared to the relationship for the phase matching condition which typically gives $0.1\,\mathrm{nm\,K^{-1}}$, the influence of the PPNC temperature on the Bragg-resonance is about three times weaker.

The reflectance spectrum obtained by varying the temperature of the PPNC corresponds to the reflectances determined by varying the wavelength of the incident light. Several local maxima of decreasing height can be distinguished near the central resonance. At reflectance levels of the order of 10^{-4}, undefined oscillations of the PPNC reflectance are observed. With respect to the relatively high reflectance level around 10^{-4} at wavelengths separated by several dozens of nanometers from the resonance, perturbed periodic poling can be assumed. With the wavelength dependence of the reflectance near the resonance, a pure duty cycle error is reasonable. While the far field intensity patterns indicate that the reflection is composed of at least two waveguide modes, a quantitative analysis of the DW reflectivity and the poling errors is not possible until all reflected waveguide modes and their relative contributions to the reflection are identified.

The occurrence of multiple waveguide modes in the reflection from the periodic poling is further investigated with the RW PPNC designed for 1178 nm. The PPNC is characterized in a wavelength range from 1174.0 nm to 1177.5 nm. In addition to the default PPNC temperature of 20 °C, the PPNC is also characterized at a temperature of 50 °C. With equation (3.26), a temperature tuning of $0.0402\,\mathrm{nm\,K^{-1}}$, determined at a wavelength of 1174.0 nm and 20 °C, is calculated and the reflectance spectrum at 50 °C is mapped to wavelengths down to 1172.7 nm. In the following, the wavelengths always correspond to the default PPNC temperature of 20 °C, so wavelengths below 1174.0 nm are calculated from the thermal wavelength shift.

In figure 3.10a the measured reflectance in a wavelength range from 1172.7 nm to 1177.5 nm is shown. In the wavelength range from 1176.4 nm to 1177.2 nm significant

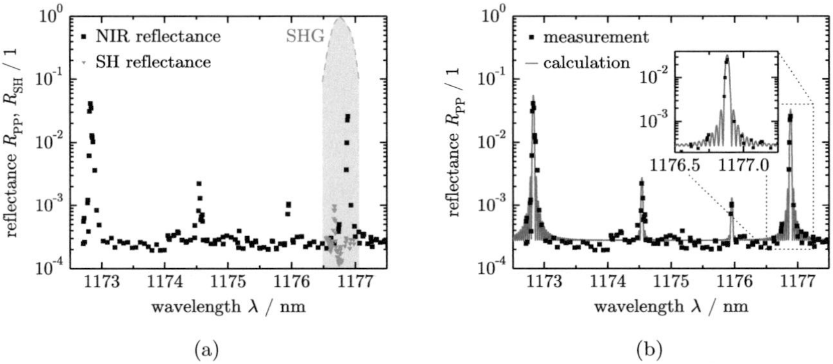

Figure 3.10.: (a) Measured reflectance of the periodic poling of the near infrared (NIR) incident light R_{PP} and the generated second harmonic (SH) light R_{SH} of a ridge waveguide periodically poled nonlinear crystal with a design wavelength of 1178 nm as a function of the wavelength of the incident light. The dashed gray line represents the normalized second harmonic generation (SHG) conversion efficiency. (b) The same measured reflectance compared to the calculation for imperfect poling with pure duty cycle error.

second harmonic (SH) emission is obtained, which is marked by the gray shaded area and the reflectance for the SH light R_{SH} is also determined. The reflectivity of the facet has been measured to be in the order of 2×10^{-3} over the entire wavelength range, which is not shown here. Four resonances can be distinguished in the reflectance spectrum of the incident near infrared (NIR) light, while one resonance is visible in the reflectance spectrum of the SH light. With respect to the far field intensity profiles presented in figure 3.11 the observed resonances in the reflectance spectrum can be assigned to individual waveguide modes (cf. figure A.1b). Starting at short wavelengths, the first three resonances are due to the reflection of the higher-order modes TE_{01}, TE_{20} and TE_{10}. The intensity profile at 1176.75 nm, where no resonance occurs, is a superposition of different waveguide modes with an asymmetric intensity distribution. The resonance at the highest wavelength is the reflection of the fundamental TE_{00} mode. Finally, the resonance in the reflectance spectrum of the SH light can be attributed to the higher-order mode TE_{10}.

The measured resonance wavelengths and maximum reflectances for the reflected

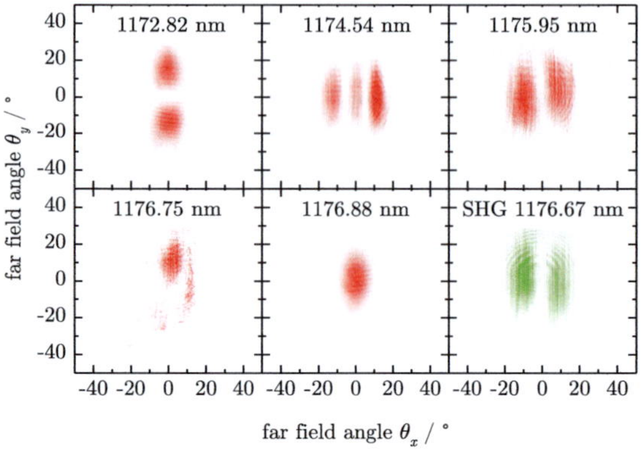

Figure 3.11.: Measured far field intensity distributions of the reflection due to the periodic poling of a ridge waveguide periodically poled nonlinear crystal with a design wavelength of 1178 nm at six different wavelengths of the incident light. The lower right color plot shows the intensity profile of the generated second harmonic light (SHG) in the visible spectral region while the corresponding wavelength of the incident light is 1176.67 nm.

modes are summarized in table 3.2. Both modes TE_{00} and TE_{01} have maximum re-

Table 3.2.: Measured resonance wavelength λ_r and maximum reflectance R_{max} of the reflected TE_ν modes of the ridge waveguide periodically poled nonlinear crystal with a design wavelength of 1178 nm. SHG TE_{10}: reflected mode corresponding to second harmonic light.

mode TE_ν	TE_{00}	TE_{10}	TE_{20}	TE_{01}	SHG TE_{10}
resonance wavelength λ_r/nm	1176.88	1175.95	1174.54	1172.82	1176.67
max. reflectance $R_{max}/10^{-2}$	2.6	0.11	0.22	4.1	0.10

flectances of the order of 10^{-2}, with the vertical higher-order mode TE_{01} having the highest reflectance. In contrast, the horizontal higher-order modes TE_{10} and TE_{20} are characterized by much smaller reflectances of the order of 10^{-3}. Between the resonances, fluctuations of the reflectance around a level of 2×10^{-4} are observed. The reflectances of the reflected SHG TE_{10} mode as well as the reflectance beyond the resonance are on the same level as the reflectances for the fundamental light wavelengths.

With the knowledge of all the reflected waveguide modes, the measured reflectance of the RW PPNC can be fitted with the theoretical model. The fit provides information about the DW amplitude reflectivity for each mode and perturbations of the poling period. In figure 3.10b the measured reflectance and the fitted reflectance based on equation (3.23) with four reflected modes as a function of the wavelength are shown. For the calculation a pure duty cycle error is assumed with a fitted standard deviation of $\sigma_{\mathrm{DE}} = 63\,\mathrm{nm}$. The fitted DW amplitude reflectivity and the effective index differences of the individual modes are given in table 3.3.

Table 3.3.: Fitted values of the domain wall (DW) amplitude reflectivity $r_{\mathrm{DW},\nu}$ and effective index difference $\Delta n_{\mathrm{eff},\nu}$ for the reflected TE_ν modes of the ridge waveguide periodically poled nonlinear crystal with a design wavelength of 1178 nm.

mode TE_ν	TE_{00}	TE_{10}	TE_{20}	TE_{01}
fitted DW amplitude reflectivity $r_{\mathrm{DW},\nu}/10^{-4}$	2.2	0.4	0.6	2.9
fitted effective index difference $\Delta n_{\mathrm{eff},\nu}/10^{-3}$	−6.7	−10.1	−15.4	−21.8

The determined DW amplitude reflectivity for the TE_{00} mode is in a reasonable range of 10^{-4}, although smaller by a factor of three compared to the values found for bulk PPNCs. Corresponding to the high reflectance maximum of the TE_{01} mode, the amplitude reflectivity for the mode conversion is even higher. Here, the vertically asymmetric geometry of the RW with the necessary material processing steps, at least for the formation of the trenches, results in the conversion to the vertical higher-order mode with antisymmetric phase distribution. In contrast, the waveguide is symmetric in the horizontal direction, which favors the conversion to the TE_{20} mode with symmetric phase distribution. The appearance of the TE_{10} mode with the lowest reflectance can be explained by waveguide inhomogeneities breaking the symmetry. For all observed modes, the fitted values agree well with calculated effective index differences given in table A.1. Thus, the observed higher-order modes are indeed converted from the incident fundamental mode, and the occurrence of higher-order modes in the pump light can be neglected. Since the optical feedback of the periodic poling of the PPNC is only a superposition of guided waveguide modes, the corresponding light is directed to the pump laser source.

Finally, the DCW PPNC with a design wavelength of 1064 nm is characterized. In figure 3.12a the measured reflectance is shown as a function of the incident wavelength. Again, the gray shaded area and dotted line visualize the wavelength region with efficient SHG, while the inset shows the reflectance in the wavelength range from 1063.9 nm to 1064.3 nm in more detail.

The reflectance is characterized by several maxima in the spectral region around 1064 nm, exceeding levels of 10^{-2} with a maximum of 3.5×10^{-2} at 1064.16 nm. In the other wavelength regions, a fluctuating reflectance is observed with an average value slightly below 10^{-3}. In contrast to the RW PPNCs investigated previously, the individual reflectance maxima can be assigned to only two waveguide modes. The exemplary far field intensity distributions at 1064.10 nm and 1064.16 nm are shown in the insets of figure 3.12b. Compared to the calculations in figure A.2b, the far field intensity distributions can be associated with the fundamental LP_{01} and higher-order LP_{11} mode. Thus, the three most intense maxima starting at 1064.16 nm can be assigned to the LP_{01} mode, while the series of local reflectance maxima with heights around 1×10^{-2} at shorter wavelengths correspond to the LP_{11} mode.

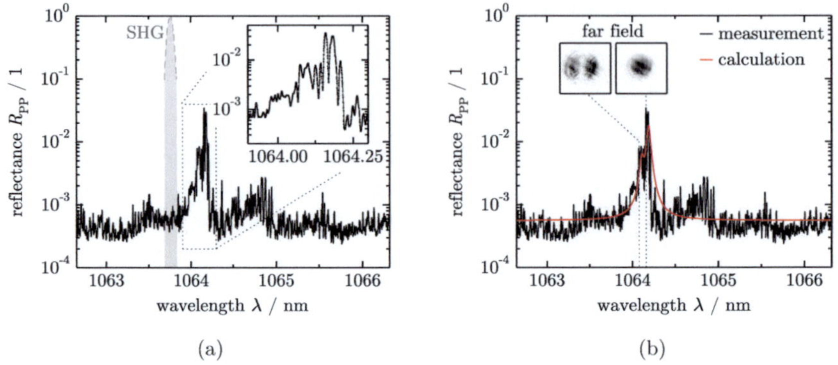

Figure 3.12.: (a) Measured reflectance of the periodic poling R_{PP} of a depressed cladding waveguide periodically poled nonlinear crystal with a design wavelength of 1064 nm as a function of the wavelength of the incident light. The dashed gray line depicts the normalized second harmonic generation (SHG) conversion efficiency. (b) The same measured reflectance compared to the calculated reflectance for imperfect poling with duty cycle and period error while the insets show the far field intensity distributions of the feedback at 1064.10 nm and 1064.16 nm.

The occurrence of multiple reflectance resonances for each mode indicates the influence of the period error type. In figure 3.12b the measured reflectance is shown in comparison with the calculation using equation (3.23) for two guided modes. The standard deviations for duty cycle and period error type σ_{DE} and σ_{PE}, as well as the effective index differences and DW amplitude reflectivities are again determined by a fit.

The fitted values for the standard deviation are $\sigma_{\mathrm{DE}} = 71.0\,\mathrm{nm}$ and $\sigma_{\mathrm{PE}} = 2.4\,\mathrm{nm}$. Thus, both error types have a significant influence on the reflectance spectrum, where the reflectance off-side the resonances is mostly affected by the duty cycle error, while the period error yields multiple reflectance maxima. However, it remains unclear whether the period error results from a disturbed poling period or from tolerances in the waveguide inscription. To confirm the period error type for periodic poling beyond doubt would require a crystal sample without waveguide structures.

In any case, the determined standard deviations are still too small to significantly affect the SHG conversion efficiency of PPNC which is still higher than 99 % of the ideal value. The DW amplitude reflectivities for the fundamental LP_{01} mode are found to be 2.5×10^{-4} which is comparable to the reflectance obtained for the LN PPNCs. With 1.2×10^{-4} the amplitude reflectivity for the conversion to the LP_{11} mode is smaller by a factor of two. Again, influences from the waveguide cladding definition processing steps can be expected. The fitted effective index differences of the LP_{01} and LP_{11} modes are -1.5×10^{-4} and -3.9×10^{-4}, respectively. The fitted values are slightly higher than the calculated effective index differences in table A.2. However, the results are reasonable taking into account the used simplified geometry and the unknown actual refractive index profile of the DCW.

3.5. Summary

In this chapter, the spectral and spatial reflection properties of periodically poled nonlinear crystals (PPNC) made of different materials and with different geometries have been investigated. In general, it could be shown that the periodic poling structure, which suffers from local refractive index variations at the domain walls (DW), acts as a Bragg-grating with typical reflectance spectra for the incident light. Characteristic resonances occur with maximum reflectances exceeding the percentage level. In addition, the reflectance spectra are strongly influenced by the manufacturing tolerances of the periodic poling and possible further processing steps, e.g. for the fabrication of waveguides. Therefore, significant reflectances of the order of 10^{-4} or higher are obtained even far from the resonances. Regarding the spatial emission characteristics, a significant part of the feedback of PPNCs is strongly directed towards the incident laser source. The actual transverse intensity distributions depend strongly on the wavelength and the geometry of the PPNC.

With the characterization of a lithium niobate (LN) bulk PPNC with a design wavelength of 1122 nm the reflection properties of the solitary periodic poling are investigated. Each DW between two opposing ferroelectric domains acts as a plane reflecting surface with an amplitude reflectivity of 6.7×10^{-4} which agrees well with values reported in the literature. It is also found that the positions of the DW are subject to duty cycle errors with random fluctuations with a standard deviation of 92 nm. According to the reflecting DW, the reflecting properties of the bulk PPNC are similar to those of weakly reflecting VBGs. Thus, characteristic reflectance spectra with peak reflectances of 7.0×10^{-3} and an average reflectance of 10^{-3} offsite of the maximum, as well as distinct spatial intensity interference patterns of the reflection are determined. The intensity patterns of the reflection are characterized by concentric interference fringes and possibly local intensity minima in the center of the pattern. Compared to the beam waist diameter of the incident Gaussian light beam, the width of the imaged near field intensity distribution in the center of the bulk PPNC varies strongly with the incident wavelength and can be smaller or larger by at least a factor of two.

For LN PPNCs with a ridge waveguide (RW) with design wavelengths of 1122 nm and 1064 nm, similar reflective properties of the periodic poling have been observed. A duty cycle error of the DW positions with a standard deviation of 63 nm and an amplitude reflectivity of 2.2×10^{-4} are determined. The material processing steps for the formation of the RW increase the refractive index perturbations of the periodic poling. Thus, the forward propagating pump light in the fundamental mode is not only reflected as it is, but also converted into higher-order waveguide modes with strong individual resonances in the reflectance spectrum. The total reflectance varies between a mean of about 10^{-4} offsite the resonances and up to 5.5×10^{-2} for the strongest resonance. According to the reflection of multiple waveguide modes, the spatial intensity patterns of the reflection correspond to the superposition of the modes.

The characterization of a lithium tantalate PPNC with a design wavelength of 1064 nm and a circular depressed cladding waveguide (DCW) also yields reflectance spectra with multiple resonances. With the obtained DW amplitude reflectivities of 2.5×10^{-4} and 1.2×10^{-4} for the fundamental LP_{01} and higher-order LP_{11} modes, the values are similar to those of LN PPNCs. In addition to a duty cycle error of the DW positions with a standard deviation of 71.0 nm, a small period error with a standard deviation of 2.4 nm has been determined. As a result, there are multiple resonance maxima for the same guided mode.

4. Optical Feedback of Semiconductor Optical Amplifiers

The optical feedback emission of semiconductor optical amplifiers (SOA) is analyzed in the following chapter. The focus is on edge-emitting traveling-wave SOAs that are typically used in high-power diode laser systems. Such SOAs are characterized by antireflection-coated input and output facets and at least planar waveguiding in the vertical epitaxial layer plane. The incident light from a seed laser source is coupled into the SOA and amplified in a single pass.

The optical feedback generated by the SOA is caused by two different effects. One contribution to the optical feedback results from reflections of the coupled light at the facets of the SOA. In particular, the reflection at the front facet results in a backward propagating wave. Here, the amplification of the backward propagating wave due to the optical gain in the SOA must also be taken into account. The second contribution to the feedback emission from SOAs comes from spontaneous emission. Spontaneous emission occurs increasingly in regions of high unsaturated optical gain [72], [73].

Both the backward propagating wave inside the SOA and the spontaneous emission can also have a negative impact on the SOA performance. Backward propagating waves can limit the optical output power [74], [75] or lead to a degradation of the spatial beam quality [76]–[79] of the output emission of the SOA. Similarly, limitations of the output performance of the SOA with respect to spontaneous emission have been reported [80], [81].

In order to optimize the SOA in terms of output performance, the causes that lead to the emission of optical feedback are thus already partially addressed. A proper antireflection coating with residual reflectivities of the order of 10^{-4} on the facets is crucial to achieve satisfactory device performance [75]. If applicable to the particular SOA design, angled facets can further reduce their effective reflectivity [82]–[84]. Even if the SOA optimizations result in the output performance no longer being affected by the backward propagating wave or spontaneous emission, the effect on the emission of optical feedback may still be relevant. Therefore, a detailed separate investigation of the optical feedback emission is required.

The feedback emission properties depend on the design and horizontal waveguiding characteristics of the SOA. In broad area amplifiers or SOAs with horizontal waveguiding structure such as ridge waveguide amplifiers (RWA), the coupled seed light does not diverge significantly. As a result, the forward and backward propagating waves have a high spatial overlap. Such configurations are similar to Fabry-Pérot diode lasers with strongly asymmetric facet reflectivities [85]. Due to the longitudinally asymmetric intensity distributions, strong longitudinal spatial hole burning occurs [86]. Consequently, the backward traveling wave experiences strong optical gain, especially where the intensity of the forward traveling wave is small, e.g. near the back facet. Even with a facet reflectivity of 1×10^{-4}, the backward emitted power can reach 1×10^{-2} of the incident seed power level [74].

Tapered amplifiers (TPA) comprise a taper section in which the forward propagating light diverges due to the lack of a waveguiding structure. Here, the backward propagating wave of a reflection at the front facet diverges further and has a greatly reduced spatial overlap with the forward propagating wave. On the one hand, the backward propagating wave is therefore less amplified. On the other hand, intensity contributions arise outside the gain regions. For tapered diode lasers, which have a similar geometry to TPAs but have typical front facet reflectivities around 1×10^{-2}, such parts of the intensity distributions can even reach the back facet [78], [79]. With decreasing facet reflectivities to the level of 1×10^{-3}, which is already close to the typical range used for TPAs, the intensity distributions outside the gain region are typically strongly attenuated [78]. While the intensity contributions due to reflections are reduced, spontaneous emission can be distinguished in the intensity distribution at the input facet of TPAs [87], [88]. However, a detailed analysis of the total emitted feedback power and the spatial emission characteristics is not yet available.

This chapter presents the theoretical and experimental investigations of the feedback emission of traveling wave SOAs with different horizontal designs. An RWA and a TPA, each with a design wavelength of 1120 nm, are examined as examples. The first section presents the theoretical description of light propagation through SOAs based on a tailored multiphysics model. In the second and third section the investigated SOA samples as well as the measurement procedure are presented in detail. Experimental and simulated results of the feedback emission characteristics are given in the fourth section, while the summary of this chapter follows in the fifth section.

4.1. Theoretical Modeling

The simulation of light propagating through semiconductor optical amplifiers (SOA) requires not only an optical model, but also models for carrier transport and heat flow to account for light-matter interactions. Since all these different effects are mutually coupled and result in nonlinear equations, numerical simulations are needed.

In general, the simulation of edge-emitting semiconductor lasers or amplifiers is a challenging task requiring sophisticated numerics with high resource demands [89]. However, the special conditions for SOAs and the restriction to steady-state considerations lead to several simplifications. Many characteristics of the SOA emission are strongly dependent on the incident seed light, which in turn is independent of the SOA. Therefore, the beam propagation method (BPM) is well suited for the simulation of SOAs and is widely used [74], [90]–[92].

In figure 4.1a the schematic of an SOA is shown. The light propagating inside the SOA is confined in the vertical waveguide. By applying a forward bias between the n-contact and the p-contact, a current is injected that is spatially defined in the horizontal-

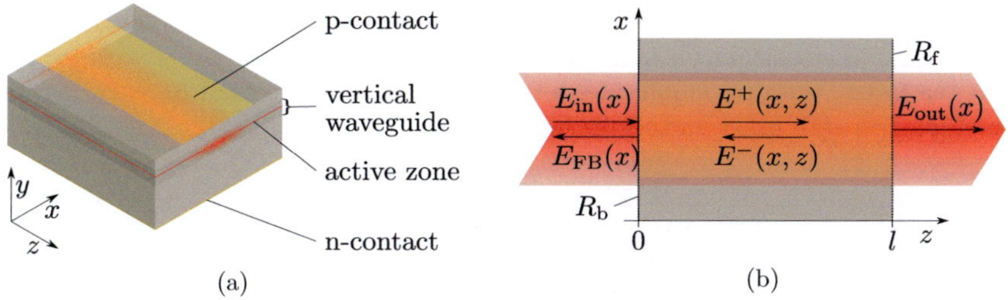

Figure 4.1.: (a) Schematic of a semiconductor optical amplifier (SOA) and (b) corresponding two-dimensional cross-section at the active zone. l: SOA length, E_{in}: electric field amplitude of incident light beam, E_{FB}: electric field amplitude of reflected feedback light beam, $E^{\pm}$: electric field amplitude of internal propagating light, E_{out}: electric field amplitude of output light beam, R_{b}: back facet reflectivity, R_{f}: front facet reflectivity.

longitudinal ($x-z$) plane by the shape of the p-contact. The resulting carrier distribution in the active zone provides the optical gain. Figure 4.1b shows a cross section of the horizontal-longitudinal plane at the vertical (y) position of the active zone. The back and front facets with reflectivities R_{b} and R_{f} are located at $z = 0$ and $z = l$, respectively, where l is the length of the SOA. The propagating light inside the SOA, the incident seed light, the feedback light and the output light are characterized by complex electric field amplitudes $E^{\pm}$, E_{in}, E_{FB} and E_{out}, respectively.

4.1.1. Multiphysics Model

The optical field and carrier distributions in the active zone can be adequately modeled by effective two-dimensional models in the horizontal-longitudinal ($x - z$) plane. In contrast, a three-dimensional model is required for the temperature distribution, whose spatial domain includes not only the SOA itself but also possible heat spreaders. The theoretical model is based on the model presented in [89], [93]–[97], with adjustments made to optimize for a steady-state BPM for SOAs.

Optical Model

The optical field inside the SOA is described by the real-valued electric field strength $\mathcal{E}$

$$\mathcal{E}(x,y,z,t) = \Re\left\{\Phi(y)[E^+(x,z)e^{-\imath\bar{n}k_0 z} + E^-(x,z)e^{\imath\bar{n}k_0 z}]e^{\imath\omega t}\right\}, \tag{4.1}$$

where E^+ and E^- are the complex slowly varying amplitudes of the forward and backward propagating waves in the ($x-z$) plane. The waves oscillate with angular frequency $\omega = c_0 k_0$ and wave vector $k_0 = 2\pi/\lambda$ with the wavelength λ. $\bar{n}$ is the reference refractive index and c_0 is the vacuum speed of light. The vertical field profile is represented by a constant real-valued function Φ which is normalized as $\int \Phi^2(y)\mathrm{d}y = 1$. This reduces the simulation of the optical field to a two-dimensional problem.

At the facets, the field amplitudes for the forward and backward propagating waves are coupled to each other and to the input field:

$$E^+(x,0) = \sqrt{1-R_\mathrm{b}}E_\mathrm{in}(x) + \sqrt{R_\mathrm{b}}E^-(x,0) \tag{4.2}$$

$$E^-(x,l) = \sqrt{R_\mathrm{f}}E^+(x,l). \tag{4.3}$$

The amplitude E_in describes the seed light that is coupled into the SOA and contributes to the forward propagating wave with the field amplitude E^+. This wave is partially reflected at the front facet, causing the backward propagating wave with the field amplitude E^-.

The field amplitudes E_out and E_FB of the forward and backward emitted light outside the SOA are defined as

$$E_\mathrm{out}(x) = \sqrt{1-R_\mathrm{f}}E^+(x,l) \tag{4.4}$$

$$E_\mathrm{FB}(x) = \sqrt{R_\mathrm{b}}E_\mathrm{in}(x) + \sqrt{1-R_\mathrm{b}}E^-(x,0). \tag{4.5}$$

Finally, the backward emitted light with field amplitude E_FB is a combination of the reflection of the seed light at the back facet and the backward propagating wave resulting from the reflection at the front facet. In this model, the field distribution at the facet inside and outside the SOA is assumed to be identical except for the constant factor of reflectivity.

The electric field amplitudes E are normalized to the optical power P

$$P = \frac{1}{2}\varepsilon_0 c_0 \int |E(x)|^2 \,\mathrm{d}x \tag{4.6}$$

with the vacuum permittivity ε_0. Thereby, the electric field amplitude are related to the measurable optical powers P_in, P_out and P_FB of the incident seed light, the output beam or the feedback beam. Furthermore, the spatial emission properties are characterized by the near and far field intensity distributions. According to the two-dimensional model for the optical field, effective vertically integrated near and far field intensity distributions are used. The vertically integrated near field intensity distribution S_x is defined by

$$S_x(x) = \frac{1}{2}\varepsilon_0 c_0 \,|E(x)|^2 \tag{4.7}$$

and the corresponding vertically integrated far field intensity is given by

$$S_\theta(\theta_x) = \frac{1}{2}\varepsilon_0 c_0 k_0 \,|\mathcal{F}\{E\}(k_0 \sin(\theta_x))|^2 . \tag{4.8}$$

Here, θ_x is the horizontal far field angle and $\mathcal{F}$ is the one-dimensional Fourier transform. The normalization ensures that integration over all horizontal far field angles yields the optical power. Far field angle dependent correction terms, which are usually used [98], are neglected following the paraxial approach.

The evolution of the field amplitudes inside the SOA obeys the paraxial wave equation

$$\pm\partial_z E^{\pm}(x,z) = -\imath\frac{1}{2\bar{n}k_0}\partial_x^2 E^{\pm}(x,z) - \imath\frac{1}{2\bar{n}k_0}\left(k_0^2 n^2(x,z,N,T) - \bar{n}^2 k_0^2\right) E^{\pm}(x,z). \quad (4.9)$$

Thereby the electric field is coupled to the local material properties via the complex refractive index $n(x,z,N,T)$.

The complex refractive index n is assumed to be piecewise constant in the z-direction and is defined by

$$n(x,z,N,T) = \bar{n} + \Delta n_0(x,z) + \Delta n_N(N) + \Delta n_T(T) + \imath\frac{g(N) - \alpha(N)}{2k_0}. \quad (4.10)$$

The contributions Δn_0, Δn_N, and Δn_T model the real-valued refractive index variations due to built-in structuring, the excess carriers N in the active zone, and the temperature distribution T. With the differential indices n'_N and n'_T for the carrier and temperature induced effective index change, the respective terms are given by

$$\Delta n_N(N) = -\sqrt{n'_N N} \quad (4.11)$$

$$\Delta n_T(T) = \int n'_T \Phi^2(y)[T(x,y,z) - T_0]\mathrm{d}y, \quad (4.12)$$

where T_0 is the reference temperature.

The imaginary part of n includes the optical gain g and losses α, which also depend on the excess carrier density:

$$g(N) = g' \ln\left(\frac{\max\{N, N_0\}}{N_{\mathrm{tr}}}\right) \quad (4.13)$$

$$\alpha(N) = \alpha_0 + f_N N. \quad (4.14)$$

Here the logarithmic gain model is applied, where g', N_0 and N_{tr} are the differential modal gain, the gain clamping carrier density and the transparency carrier density, respectively. Optical losses are modeled by the background modal absorption α_0 and the free carrier induced absorption in the active zone, which is determined by the effective absorption cross section f_N.

Carrier Model

The carrier distribution in the active zone is modeled by the vertically averaged excess carrier density N. It is defined by the steady-state diffusion equation

$$-\partial_x\frac{j_{\mathrm{dif}}(N,T)}{e_0} + \frac{j(x,z,N)}{e_0 d_{\mathrm{AZ}}} - R_{\mathrm{spon}}(N) - R_{\mathrm{stim}}(N, E^{\pm}) = 0 \quad (4.15)$$

which is subject to horizontal Neumann boundary conditions while neglecting diffusion in the longitudinal (z) direction. j_{dif} and j are the diffusion and injection current densities, whereas d_{AZ} is the thickness of the active zone and e_0 is the elementary charge.

Spontaneous and stimulated recombination processes are described by the rates R_{spon} and R_{stim}, respectively.

The currents in equation (4.15) are defined by

$$j_{\text{dif}}(N,T) = -\sigma_{\text{p}}(N)\partial_N \varphi_{\text{F}}(N,T)\partial_x N \tag{4.16}$$

$$j(x,z,N,T) = \begin{cases} \dfrac{U-\varphi_{\text{F}}(N,T)}{r_{\text{s}}} & (x,z) \in \text{p-contact} \\ \dfrac{\partial_x^2 \varphi_{\text{F}}(N,T)}{\Omega_{\text{s}}} & \text{else} \end{cases} \tag{4.17}$$

σ_{p} is the conductivity for holes and φ_{F} is the Fermi potential. The injection current j is derived from an approximate solution of the Laplace equation in the p-doped region, also referred to as the Joyce model [99]. Here, U is the bias voltage at the p-contact and $r_{\text{s}} = \int_{\text{p-layers}}(\sigma(y))^{-1}\text{d}y$ and $\Omega_{\text{s}} = (\int_{\text{p-layers}} \sigma(y)\text{d}y)^{-1}$ are the series and sheet resistances calculated from the hole conductivity in the respective p-doped layers between active zone and p-contact. As a consequence of the horizontal structuring of the SOA, including implantation or etching beyond the p-contact, the sheet resistance can vary. The hole conductivity in the active zone is given by

$$\sigma_{\text{p}}(N) = e_0(N + p_0)\mu_{\text{p}}, \tag{4.18}$$

with the equilibrium hole density p_0 and the hole mobility μ_{p}. For the calculation of the Fermi potential φ_F the *Unger* approximation [100] is used:

$$\begin{aligned} \varphi_{\text{F}}(N,T) = \frac{E_{\text{g}}}{e_0} &+ \frac{k_{\text{B}}T}{e_0}\ln\left(\exp\left[\frac{1}{0.3}\left\{\sqrt{1+0.6\frac{N+n_0}{N_{\text{c}}}}-1\right\}\right]-1\right) \\ &+ \frac{k_{\text{B}}T}{e_0}\ln\left(\exp\left[\frac{1}{0.3}\left\{\sqrt{1+0.6\frac{N+p_0}{N_{\text{v}}}}-1\right\}\right]-1\right). \end{aligned} \tag{4.19}$$

E_{g} is the bandgap energy, T is the temperature in the active zone, and k_{B} is the Boltzmann constant. Complementary to p_0, the parameter n_0 is the equilibrium carrier density for electrons, while N_{c} and N_{v} are the density of states in the conduction and valence band, respectively.

The equations for the spontaneous and stimulated rates are

$$R_{\text{spon}}(N) = A_{\text{SRH}}N + B_{\text{SE}}N^2 + C_{\text{Aug}}N^3 \tag{4.20}$$

$$R_{\text{stim}}(N,E^{\pm}) = \frac{g(N)\varepsilon_0 c_0}{2\hbar\omega d_{\text{AZ}}}\left(|E^+|^2 + |E^-|^2\right) \tag{4.21}$$

where the former couples to the electric field amplitudes of the optical field. $\hbar$ is the reduced Planck constant and A_{SRH}, B_{SE}, and C_{Aug} are the coefficients for Shockley Read Hall, spontaneous emission and Auger recombination, respectively.

Thermal Model

The temperature distribution in the entire SOA, including possibly mounted heat spreaders, is obtained from the steady-state heat equation. Similar to the calculation of excess carriers N, the longitudinal heat flow is neglected:

$$[\partial_x \kappa(y) \partial_x + \partial_y \kappa(y) \partial_y] \, T(x, y, z) = -H(x, y, z, N, T, E^{\pm}). \tag{4.22}$$

In general, the SOA can be structured in the horizontal direction, e.g. due to etching. Since the structuring is typically located only in the thin p-doped layers above the active zone, the influence on the heat flow is neglected. Therefore, the geometry of the SOA is assumed planar and the thermal conductivity κ varies only in the vertical direction. Equation (4.22) obeys boundary conditions that are defined either as Dirichlet conditions for boundaries located at a heat sink or as Neumann conditions. The source term H is modeled by

$$\begin{aligned} &H(x, y, z, N, T, E^{\pm}) = \\ &\qquad [q_{\mathrm{J}}(x, z, N, T) + q_{\mathrm{abs}}(N, E^{\pm}) + q_{\mathrm{rec}}(N, T) + q_{\mathrm{def}}(N, T, E^{\pm})] \, \delta(y - y_{\mathrm{AZ}}), \end{aligned} \tag{4.23}$$

which assumes that all heat sources are located at the vertical position of the active zone y_{AZ}. The four heat flux contributions account for the Joule heat q_{J}, the absorption heat q_{abs}, the recombination heat q_{rec}, and the defect heat q_{def}, each defined as

$$q_{\mathrm{J}}(x, z, N, T) = r_{\mathrm{s}} j^2(x, z, N, T) + \frac{d_{\mathrm{AZ}}}{\sigma_{\mathrm{p}}(N)} j_{\mathrm{dif}}^2(N, T) \tag{4.24}$$

$$q_{\mathrm{abs}}(N, E^{\pm}) = \alpha(N) \frac{\varepsilon_0 c_0}{2} \left(|E^+|^2 + |E^-|^2 \right) \tag{4.25}$$

$$q_{\mathrm{rec}}(N, T) = d_{\mathrm{AZ}} e_0 \varphi_{\mathrm{F}}(N, T) R_{\mathrm{spon}}(N) \tag{4.26}$$

$$q_{\mathrm{def}}(N, T, E^{\pm}) = d_{\mathrm{AZ}} \left[e_0 \varphi_{\mathrm{F}}(N, T) - \hbar\omega \right] R_{\mathrm{stim}}(N, E^{\pm}). \tag{4.27}$$

Although the Joule heat and especially the absorption heat are generated in an extended vertical region of the size of the p-doped region or the vertical mode profile, these distributions are approximated as a sheet heat flux. As shown in section A.6.1, this approximation is well suited and leads to only small differences to a three-dimensional distribution of heat sources.

While the geometry of the SOA in the horizontal-vertical plane $(x - y)$ is rectangular, the application of additional heat spreaders that exceed the width of the SOA can lead to more complex geometries. However, in most applications the effects of deviations from a simple rectangular geometry can be neglected and a simplified geometry is sufficient for the calculation. Further details are also given in section A.6.1.

4.1.2. Estimation of Optical Feedback Power

The full spatially-resolved simulation of the light propagating through the SOA with the multiphysics model requires several material parameters and computational effort.

If only a rough estimate of the total emitted optical power of the feedback is required, a simplified one-dimensional model can be used. The approach is similar to the model of a Fabry-Pérot laser as described in [101].

From the equations (4.2) to (4.9) simplified relations for the optical power can be derived. The gain and loss are described by effective spatially averaged quantities g_{eff} and α_{eff}, while gain saturation and longitudinal spatial hole burning are neglected. The differential equations for the power $P^{\pm}$ of the internally propagating waves then reduces to

$$\partial_z P^{\pm}(z) = \pm(g_{\text{eff}} - \alpha_{\text{eff}})P^{\pm}(z). \tag{4.28}$$

Wit respect to the typically low facet reflectivities and the coupled seed power at the back facet, the boundary conditions (4.2) and (4.3) can be simplified. First, the facet reflectivities are neglected in transmission direction. Second, the seed power P_{in} is assumed to be much stronger than the internally reflected power $R_{\text{b}}P^{-}(0)$ at the back facet and the reflection is also neglected:

$$P^{+}(0) = P_{\text{in}} \tag{4.29}$$

$$P^{-}(l) = R_{\text{f, eff}}P^{+}(l). \tag{4.30}$$

Here, the reflectivity of the front facet is replaced by an effective reflectivity $R_{\text{f, eff}}$, which includes possible losses due to reduced overlap of the forward and backward propagating waves, e.g. in tapered amplifiers. Consequently, the equations (4.4) and (4.5) transform to

$$P_{\text{out}} = P^{+}(l) \tag{4.31}$$

$$P_{\text{FB}} = R_{\text{b}}P_{\text{in}} + P^{-}(0). \tag{4.32}$$

P_{out} and P_{FB} are the output and feedback power, respectively. The different feedback contributions in equation (4.5) originating from the reflection of the seed light at the back facet and the power of the backwards propagating wave are added as powers.

The differential equation (4.28) can be solved analytically which yields

$$P^{\pm}(l) = P^{\pm}(0)e^{\pm(g_{\text{eff}} - \alpha_{\text{eff}})l}. \tag{4.33}$$

From the boundary condition (4.29) and equation (4.31) follows that the exponential in equation (4.33) can be expressed as $e^{(g_{\text{eff}} - \alpha_{\text{eff}})l} = P_{\text{out}}/P_{\text{in}}$. Inserting into equation (4.32) and using equation (4.32) gives finally

$$P_{\text{FB}} = R_{\text{b}}P_{\text{in}} + R_{\text{f, eff}}\frac{P_{\text{out}}^2}{P_{\text{in}}}. \tag{4.34}$$

With equation (4.34) the calculation of the feedback power emitted from arbitrary SOA designs is possible. All input parameters are directly accessible and no knowledge of internal material properties is required. Furthermore, the scaling properties of the input properties are clearly visible. The feedback power depends linearly on the reflectivity of both the front and the back facets, in agreement with the results of the simulations in [74]. In contrast, the feedback power depends quadratically on the output power. Only the influence of the seed power is more complicated and depends on the actual working point of the SOA.

4.2. Investigated Semiconductor Optical Amplifier Samples

The semiconductor optical amplifier (SOA) samples investigated in this work are based on two different designs. First, a ridge waveguide amplifier (RWA), where the propagating light is guided through the entire device. Second, a tapered amplifier (TPA), which has different sections with and without horizontal guiding of the propagating light.

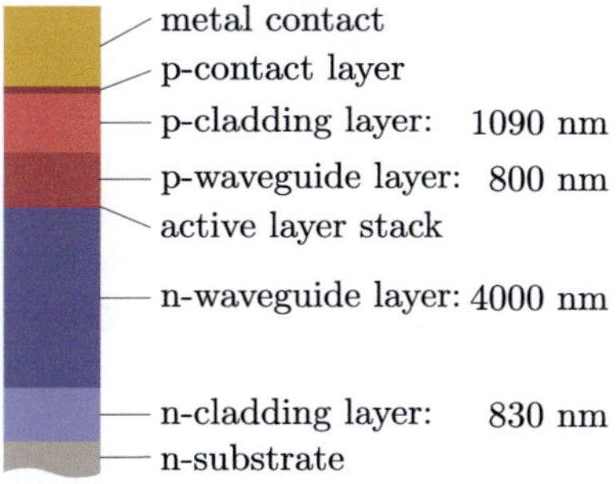

Figure 4.2.: Schematic of the asymmetric super large optical cavity vertical structure and the thickness of the main layers.

Both SOA samples share the same asymmetric super large optical cavity (ASLOC) vertical structure with a design wavelength of 1120 nm. The SOAs are fabricated by metal-organic vapor phase epitaxy on n-doped gallium arsenide substrates. Further information on the structure and its processing can be found in [102]. A schematic of the ASLOC structure with the main layers is shown in figure 4.2.

The active layer stack consists a double quantum well with a total effective thickness of 10 nm and is asymmetrically embedded in a 4.8 µm thick waveguide which is surrounded by cladding layers. A detailed list of the individual layers and corresponding material parameters is given in table B.3.

4.2.1. Ridge Wavegudie Amplifier

The RWA used for the investigation of the feedback emission is shown schematically in figure 4.3. The RWA has a length of $l = 4\,\mathrm{mm}$, a width of $w = 0.5\,\mathrm{mm}$ and a total height of $h = 0.154\,\mathrm{mm}$. The ridge waveguide (RW) with a width of $w_{\mathrm{RW}} = 4\,\mu\mathrm{m}$ extends over the entire length of the RWA and is formed by two etched trenches with a width of $w_{\mathrm{t}} = 20\,\mu\mathrm{m}$ each. This creates a refractive index step of 5×10^{-3}. Both facets are anti-reflective coated with residual reflectivities on the back and front facets determined to be $R_{\mathrm{b}} = 5 \times 10^{-4}$ and $R_{\mathrm{f}} = 3 \times 10^{-4}$, respectively. In order to further reduce the effective facet reflectivity at the back facet, the RW has an outcoupling bend within the last 1 mm, resulting in a tilt angle of $\theta_{\mathrm{RW}} = 3°$ at the back facet. Due to the bent waveguide, the RW is horizontally off-center by 33.3 µm at the back facet. The

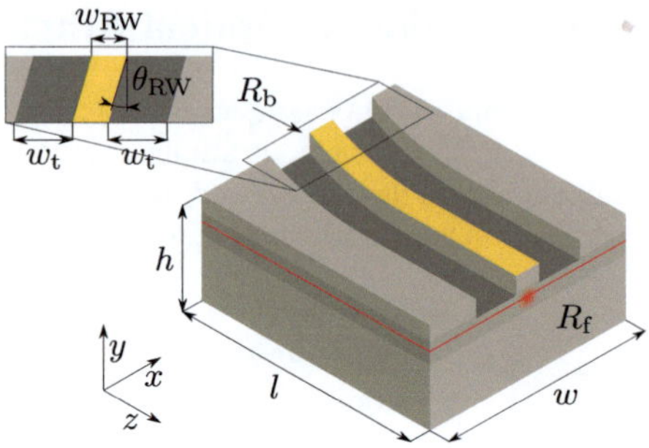

Figure 4.3.: Schematic of a ridge waveguide amplifier (RWA). l: RWA length, w: RWA width, h: RWA height, w_{RW}: ridge waveguide (RW) width, w_t: etched trench width, θ_{RW}: RW angle relative to back facet, R_b: back facet reflectivity, R_f: front facet reflectivity.

p-contact is highlighted by the yellow area in figure 4.3. Due to the etched trenches and the application of insulating layers, the p-contact shares the same area as the RW.

The RWA is soldered with the epitaxial side down to a copper-tungsten heat spreader. Operated at room temperature, the RWA is designed for output power up to the watt-level. The required seed power level is about 50 mW, while typical injection currents are in the range of 1 A and 2 A. A similar 4 mm long RWA is part of a monolithic MOPA and has been characterized in [103] where output powers close to 600 mW have been achieved at a maximum current of 1 A. At a slightly different emission wavelength of 1064 nm, a comparable RWA offers more than 1 W optical output power in a hybrid MOPA configuration [84]. Compared to common RWA designs with a straight but slanted RW [84], the design of the RWA used in this work is particular. On the one hand, the amplified output beam passes the front facet perpendicularly, which is similar to TPAs and allows a better comparability of the respective results. On the other hand, the bent waveguide is necessary to suppress Fabry-Pérot modes by reducing the effective reflectivity of at least the back facet.

According to equation (4.34) and the respective reflectivities of the front and back facets, the feedback power can be estimated. At a working point with a seed power of 50 mW and 1 W optical output power, the power of the feedback is expected to be around 6 mW. With respect to the seed power, this corresponds to an effective reflectance of the RWA of 1.2×10^{-1}.

4.2.2. Tapered Amplifier

The investigated TPA consists of a RW section followed by tapered (TP) section without horizontal waveguiding. A schematic of the TPA is shown in figure 4.4. The RW section has a length of $l_{RW} = 2$ mm and the TP section is $l_{TP} = 4$ mm long, giving a total TPA length of 6 mm. The TPA has a width of $w = 0.8$ mm and a height of $h = 0.154$ mm. The

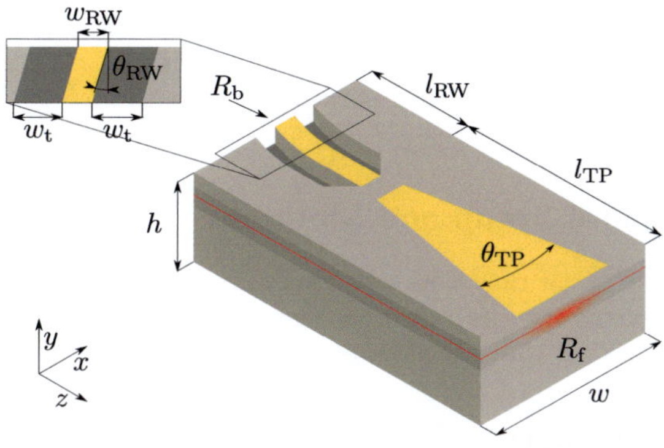

Figure 4.4.: Schematic of a tapered amplifier (TPA) consisting of a ridge waveguide (RW) and a tapered (TP) section. l_{RW}: RW section length, l_{TP}: TP section length, w: TPA width, h: TPA height, w_{RW}: RW width, w_t: etched trench width, θ_{RW}: RW angle relative to back facet, θ_{TP}: TP opening angle, R_b: back facet reflectivity, R_f: front facet reflectivity.

RW is defined by two $w_t = 48\,\mu m$ wide etched trenches and has a width of $w_{RW} = 4\,\mu m$. Similar to the RWA, the RW has an outcoupling bend towards the back facet. However, the bend is realized within $0.35\,\mu m$, resulting in a stronger curvature. The RW has an angle of $\theta_{RW} = 3°$ relative to the back facet, while the center of the RW has a horizontal offset of $11.7\,\mu m$. The opening angle of $\theta_{TP} = 6°$ of the TP section matches the divergence angle of the light from the RW section. The front and back facets are antireflection coated, with the residual reflectivity determined to be $R_f = 3 \times 10^{-4}$ and $R_b = 1 \times 10^{-4}$, respectively. The p-contact of the RW section is formed similarly to the RWA by etching and insulating layers. In contrast, the p-contact of the TP section is defined by ion implantation.

To achieve independent current injection in the RW and TP sections, the TPA is soldered with the epitaxial side up to a copper-tungsten heat spreader. The TP section is equipped with an additional $250\,\mu m$ thick copper-tungsten heat spreader (not shown in figure 4.4) that is soldered on top and covers the entire section. Thereby homogenization of the temperature distribution and a better heat extraction is achieved. Similar to the RWA, the TPA is designed for operation at room temperature. A typical working point for the TPA is a seed power of 50 mW and an RW current of 600 mA. The current through the TP section is adjusted to achieve the desired power level while currents exceeding 10 A are possible. TPAs with the same geometric layout but without bent waveguide have been used in [27] to generate nearly 9 W optical output power at a TP current as high as 14 A.

At moderate working points with output powers of 2 W to 5 W, the optical feedback power is estimated by equation (4.34) to be in the range of 1.6 mW to 10 mW, yielding effective reflectances of the TPA between 4×10^{-2} and 2×10^{-1}. Here, the reflectivity of the front facet is multiplied by the coupling efficiency of the backpropagating light into the RW section. Following the geometric beam expansion in the TP section and evaluating the overlap integral, the coupling efficiency into the RW section is approximately 0.054.

4.3. Experimental Setup and Methods

The experimental setup for the characterization of the semiconductor optical amplifiers (SOA) is comparable to the setup used for the periodically poled nonlinear crystals in section 3.3. In figure 4.5 the schematic of the setup is shown. A distributed Bragg-

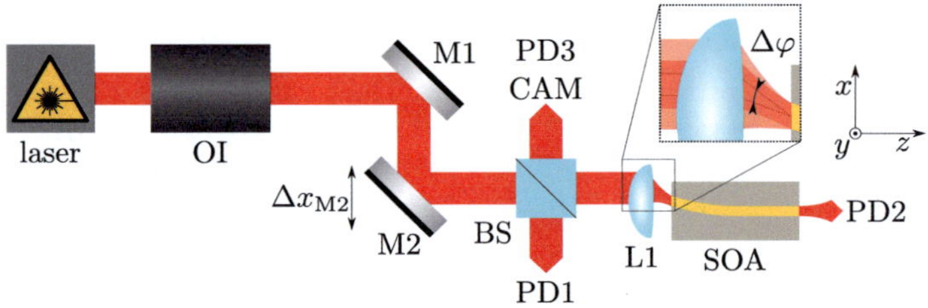

Figure 4.5.: Schematic of the experimental setup for the characterization of semiconductor optical amplifiers (SOA). OI: optical isolator, L1: aspherical lens, M1 / M2: plane mirrors, BS: beamsplitter, PD1 - PD3: power detectors, CAM: CCD camera, Δx_{M2}: horizontal displacement of M2, $\Delta\varphi$: variation of the incidence angle due to variation of Δx_{M2}.

reflector ridge waveguide laser with an emission wavelength of 1120 nm is used as the laser source, providing the seed light as a collimated beam. More information about the laser can be found in section C.1. The light passes through an optical isolator (OI) which has an isolation of at least 40 dB. Two silver mirrors M1 and M2 direct the light through a beamsplitter (BS) to the aspherical coupling lens L1 with a focal length of 8 mm. The reflected beam from the BS is directed to the power detector PD1 (*XLP12-3S-H2*) to determine the input seed power.

The mirror M2 is movable in the x-direction, which causes a slight offset of the beam with respect to the center of the lens L1 and finally results in a defined angle of incidence into the SOA. By a movement of Δx_{M2} the angle of incidence varies by $\Delta\varphi = \arctan(\Delta x_{M2}/f_{L1})$, where f_{L1} is the focal length of L1. Thereby ideal coupling or a defined misalignment of the seed light into the RW of the SOA can be realized. The SOA on its heat spreader is mounted on a heat sink which is actively controlled to a defined temperature of 25 °C. The optical output power of the SOA is measured with the power detector PD2 (*UP19K-15S-W5*).

The feedback emission from the SOA is first collimated by the lens L1 and partially reflected by the BS. With the power detector PD3 (*PH20-Ge-OD2*) and the camera (CAM) *Scorpion SCOR-20-SOM* the optical power and spatial intensity distributions of the SOA feedback are characterized. For the near field intensity distribution, the SOA back facet is imaged. The measurement procedure and the used lenses are described in section C.4.

The setup is adjusted using the output power P_{out} of the SOA, which is directly obtained from PD2, as the optimization criterion. The vertical epitaxial structure and the horizontal ridge waveguide of the seed laser source are identical to those of the SOA. Thus, an ideal coupling of the seed power into the SOA can be assumed. From the measured power P_{PD1} with power detector PD1 and the reflectivity R_{BS} of the beamsplitter BS, the seed power P_{in} in front of the SOA is obtained by

$$P_{\mathrm{in}} = \frac{1 - R_{\mathrm{BS}}}{R_{\mathrm{BS}}} P_{\mathrm{PD1}}. \tag{4.35}$$

With the seed laser source used, seed powers of up to 300 mW can be achieved. The optical power P_{FB} of the feedback of the SOA is calculated from the measured power P_{PD3}, obtained by power detector PD3:

$$P_{\mathrm{FB}} = \frac{1}{R_{\mathrm{BS}}} P_{\mathrm{PD3}}. \tag{4.36}$$

In addition to the optical power, the feedback is also described by an effective reflectance R_{SOA}, defined by the ratio of the feedback power P_{FB} to the seed power P_{in}:

$$R_{\mathrm{SOA}} = \frac{1}{1 - R_{\mathrm{BS}}} \frac{P_{\mathrm{PD3}}}{P_{\mathrm{PD1}}}. \tag{4.37}$$

Since the SOA is an active device that can emit optical power even without incident seed light, the reflectance R_{SOA} is not limited to values up to 1.

For this work, a custom implementation of the multiphysics model presented in section 4.1.1 has been realized using *Python*. More information about the computation scheme can be found in section A.6. The used material parameters for the vertical layer structure are provided in table B.3, while specific parameters for the ridge waveguide amplifier and the tapered amplifier can be found in tables B.4 and B.5, respectively. For both SOAs, the input electric field is modeled as a Gaussian profile, corresponding to a beam diameter of 4 µm and centered on the center of the ridge waveguide at the back facet.

4.4. Results

The optical feedback emission of the ridge waveguide amplifier (RWA) and the tapered amplifier (TPA) is investigated under various operating conditions. The experimental results are complemented by results of simulations based on the multiphysics model to further characterize the feedback emission behavior. Both the optical power of the feedback and its spatial emission characteristics are studied.

4.4.1. Optical Feedback of Ridge Waveguide Amplifiers

The optical feedback of the RWA is first characterized in terms of the emitted power. In figure 4.6a the optical output power and the feedback power are shown as a function of the injection current. The RWA is operated at injection currents up to 1.75 A and

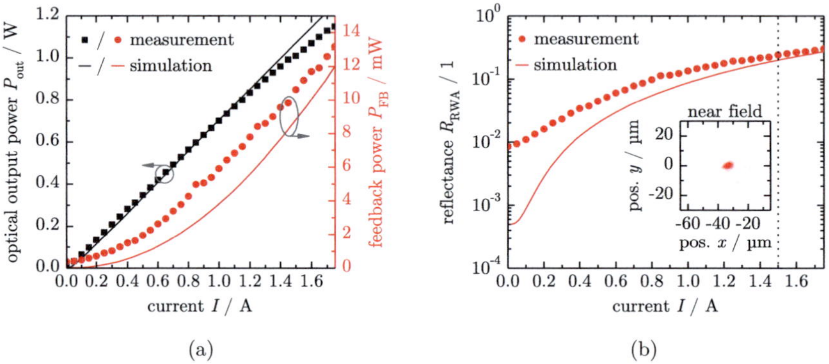

Figure 4.6.: (a) Output power and feedback power of a ridge waveguide amplifier (RWA) as a function of the current at a fixed seed power of $P_{\text{in}} = 44\,\text{mW}$. (b) Reflectance of the RWA with respect to the seed power as a function of the current. The inset shows the near field intensity distribution at the back facet at a working point of $I = 1.5\,\text{A}$.

a maximum optical output power of 1.15 W is achieved. The feedback power reaches a maximum level of 13.2 mW at the highest injection current. While the output power increases almost linearly with the current, the feedback power shows a nonlinear behavior.

In figure 4.6b the reflectance of the RWA is shown as a function of the current. The measured reflectance ranges from nearly 10^{-2} to a maximum of 3.0×10^{-1}. The inset of figure 4.6b displays the near field intensity distribution at the back facet at an injection current of 1.5 A. According to the intensity distribution, the feedback is almost entirely emitted from the RW.

The simulated output and feedback power agree with the measured results, although the feedback power is consistently underestimated. Especially in the logarithmic scale in figure 4.6b the differences between measurement and simulation become more visible. While a good agreement is obtained at high currents, the measured and simulated RWA reflectance differ by more than an order of magnitude at 0 A. The reflectivity of the facets of 5×10^{-4} and 3×10^{-4} cannot explain the high measured reflectance of the passive RWA. Possible reasons for the increased reflectance are mismatches of incident seed light at the back facet of the RWA. Since the antireflection coating is optimized for the waveguide mode in the RW with corresponding effective index, deviating beam contributions may exhibit higher reflectances. In addition, the metal contact at the top of the RWA strongly reflects misaligned beam contributions. For non-vanishing injection currents, spontaneous emission can also increase the feedback emission and consequently the reflectance of the RWA.

Next, the effect of tolerances in the coupling angle of the incident seed light is investigated. At a working point of 1.5 A, the horizontal coupling angle φ of the seed light is varied between 0° and 20° relative to the back facet. Thus, $\varphi = 0°$ corresponds to normal incidence at the back facet. To evaluate the influence of the coupling on the output power and the feedback power, both are expressed as a relative power P_{rel}. The relative powers are defined as the ratio of the output power P_{out} or feedback power P_{FB} to the power level $P_{\mathrm{out},10°}$ and $P_{\mathrm{FB},10°}$ at the ideal coupling angle of $\varphi = 10°$. In figure 4.7a the relative powers as a function of the coupling angle are shown. While the seed light is

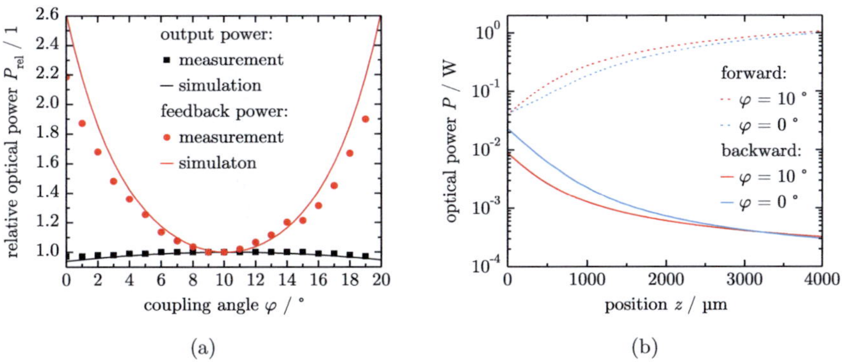

Figure 4.7.: (a) Relative optical power of the output power and feedback power of a ridge waveguide amplifier (RWA) at a fixed current of $I = 1.5$ A and a seed power of $P_{\mathrm{in}} = 44$ mW but varying coupling angle φ. The relative optical power is defined as the ratio of the output power P_{out} or feedback power P_{FB} to the power levels $P_{\mathrm{out},10°}$ or $P_{\mathrm{FB},10°}$ at a coupling angle of $\varphi = 10°$. While ideal coupling into the bent waveguide is achieved at $\varphi = 10°$, $\varphi = 0°$ corresponds to normal incidence at the facet. (b) Optical power of the forward and backward propagating waves inside the RWA as a function of the longitudinal position at the same working point and two coupling angles.

ideally coupled into the bent RW at an angle of 10°, a coupling angle of 10° corresponds to normal incidence on the back facet.

The output power of the RWA is only slightly affected by the coupling angle. Even at the maximum misaligned coupling angles of 0° or 20°, there is a small power decrease of 3 %. In contrast, the feedback power increases strongly and almost doubles. The simulation results agree with the experimental data, although the influence of the coupling angle is slightly overestimated for both the output and the feedback power. In order to explain the strong increase of the feedback power and the almost unaffected output power, the longitudinal power distribution provided by the simulation is considered in detail. The optical power of the forward and backward propagating waves as a function of the longitudinal position at an injection current of 1.5 A and two different coupling angles is shown in figure 4.7b. Due to the large power difference between the seed power and the output power, strong longitudinal spatial hole burning is present. At the ideal coupling angle of 10°, the incident seed power is strongly amplified within the first 1 mm, then the amplification saturates and power increases moderately in the remaining RWA. In case of misaligned coupling with an angle of 0° the power increase at the beginning of the RWA is significantly lower. Here, less power is coupled to the RW and the saturation of the amplification is less pronounced, occurring in the range of $z = 1.5\,\mathrm{mm}$ to $z = 2\,\mathrm{mm}$. Consistent with the low output power reduction at $\varphi = 0°$, the power of the forward propagating wave also reaches the watt level at the end of the RWA.

Consequently, the powers of the backward propagating waves near the front facet are comparable. The power increases toward the back facet with the strongest growth near the facet. Due to the reduced power of the forward propagating wave for misaligned coupling, more optical gain is available for the backward propagating wave, especially near the back facet. Thus, more optical feedback is emitted for $\varphi = 0°$.

Finally, the spatial emission characteristics of the feedback light are examined based on the vertically integrated near and far field intensity distributions. In figure 4.8 the respective intensity distributions obtained from measurements and simulations at a fixed injection current of 1.5 A and two coupling angles are shown. The measured near field

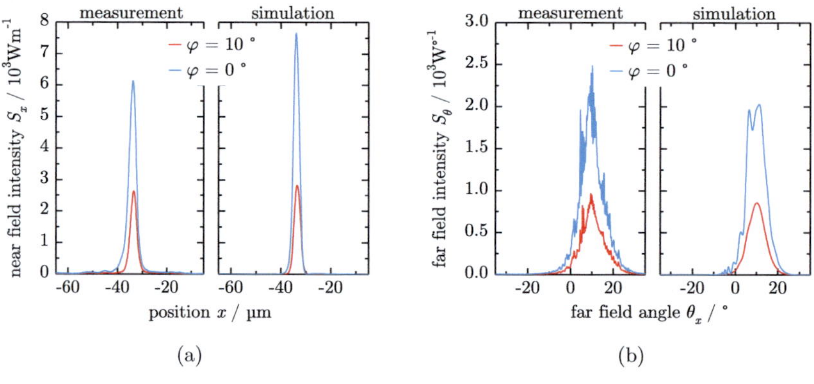

Figure 4.8.: Vertically integrated (a) near field and (b) far field intensity distributions of the feedback emission of a ridge waveguide amplifier (RWA) for two different coupling angles φ of the seed light. While ideal coupling into the bent waveguide is achieved at $\varphi = 10°$, $\varphi = 0°$ corresponds to normal incidence at the facet. The RWA is operated at a current of $I = 1.5\,\mathrm{A}$ and a seed power of $P_{\mathrm{in}} = 44\,\mathrm{mW}$.

intensity distribution, shown in figure 4.8a, is dominated by a single almost Gaussian shaped lobe resulting from the emission of the RW. Especially when compared to the two-dimensional near-field distribution in figure 4.6b, very small side lobes are present at positions around $-20\,\mu m$. When the seed light coupled perpendicular to the back facet ($\varphi = 0°$), the maximum intensity doubles and the small side lobes become more visible. Furthermore, a small spatial shift of the position of the maximum intensity from $-33.3\,\mu m$ to $-33.9\,\mu m$ is observed. The simulation results yield the same characteristics of the near field intensity distribution. In the case of misaligned coupling, the increase in the maximum intensity is consistent with the increase in total feedback power, while the occurrence of side lobes is less pronounced compared to the measurement.

In figure 4.8b the far field intensity distributions are shown. Both the measured and the simulated far field intensity distributions yield main far field angles around $\theta_x = 10°$. Thus, the backward propagating light is guided in the RW and emitted accordingly, regardless of the coupling angle. The measured far field intensity distribution suffers from several modulations, which are most reasonably a measurement artifact. The envelope of the intensity distribution is characterized by a slight asymmetry with a shoulder at $\theta_x \approx 5°$, which is consistent with the simulation results.

In contrast to the emission from the RW, originating from a reflection at the front facet, the emission direction of the reflection at the back facet depends on the coupling angle. Since the feedback power of the back facet reflection is much smaller than the total feedback power, its contribution cannot be resolved in the data. However, for the passive RWA, a measurement of the far field intensity distribution shows that the reflection at the back facet follows the law of reflection. Although the reflectance of the passive RWA is increased by different effects, a major part of the reflection is deflected to far field angles around $-10°$ in the case of ideal coupling.

The near field and the far field intensity distributions indicate that a large fraction of the feedback is emitted from the RW region with an emission angle spectrum associated with the light guiding properties of the RW. In contrast, the emission of undirected spontaneous emission, de-localized from the RW and with a broad angular spectrum, is not significantly observed. Therefore, the contribution of spontaneous emission to the total feedback emission is small for the investigated RWA.

4.4.2. Optical Feedback of Tapered Amplifiers

First, the dependence of the output and the feedback power on the seed power is investigated. The measurement and simulation results are given in figure 4.9. At seed power levels below 10 mW, a strong increase of the output power from 45 mW to more than 1.5 W is observed. At higher seed power levels the TPA is in the saturation regime and the output power increases only moderately. The maximum output power of 1.8 W is reached at the highest seed power of 49 mW. Complementary to the output power, the feedback power first decreases sharply with the seed power, followed by only a small decrease at seed powers greater than 10 mW. The simulation results show the same dependence, although high output power levels are reached at much lower seed power levels. While the simulated output power is in agreement with the measured results at high seed power levels, the feedback power is almost always underestimated by a factor

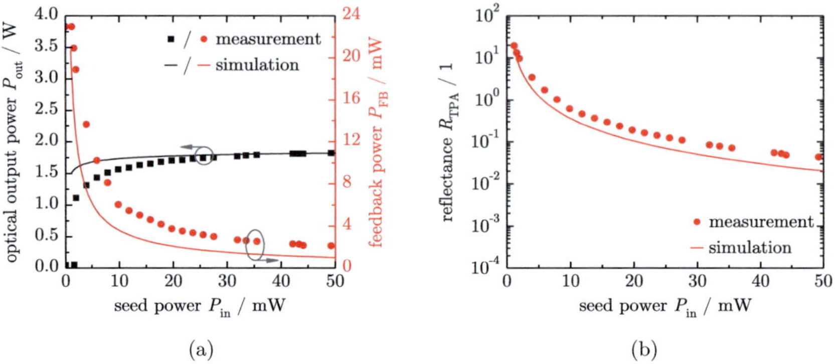

Figure 4.9.: (a) Output power and feedback power of a tapered amplifier (TPA) as a function of the seed power at ridge waveguide (RW) and taper (TP) currents of $I_{RW} = 600\,mA$ and $I_{TP} = 4\,A$. (b) Corresponding reflectance of the TPA with respect to the seed power as a function of seed power.

of two.

The reflectance of the TPA, shown in figure 4.9b, decreases from about 20 down to 4.3×10^{-2}. Since the TPA reflectance is related to the seed power, a reduction in TPA reflectance is obtained even if the feedback power remains constant. The combination of seed power increase and feedback power reduction attenuates the reflectance by an order of magnitude in the range from 10 mW to 50 mW.

Next, the influence of the RW and TP injection currents on the output and feedback emission of the TPA is considered. Figure 4.10a shows the results for varying TP current are shown. The optical output power increases from initially 22.0 mW to 3.7 W at the highest TP current of 7 A, while a linear dependence is observed for currents greater than 1.5 A. The power of the optical feedback reaches a maximum of 5.8 mW. Again, the nearly quadratic dependence of the feedback power on the output power and the feedback power levels agree also with the estimates.

The simulation can reproduce the emission behavior of the TPA in terms of output power. Good agreement is also obtained for the simulated feedback power for working points with high TP currents and output powers, respectively. However, the feedback power is underestimated especially at low TP currents, where the simulations yield almost vanishing feedback power. In contrast, the measured feedback power is not less than 0.60 mW. In figure 4.10b the corresponding reflectance of the TPA as a function of the TP current is given. Additionally, the reflectance of the passive TPA, with no current applied to the RW and TP sections at all, is given as a reference. The measured reflectance ranges from 1.4×10^{-2} to 1.3×10^{-1}. With respect to the nearly vanishing output power at low TP currents, the contribution of reflections at the front facet is negligible. Consequently, the simulated reflectance approaches 1×10^{-4}, the reflectivity

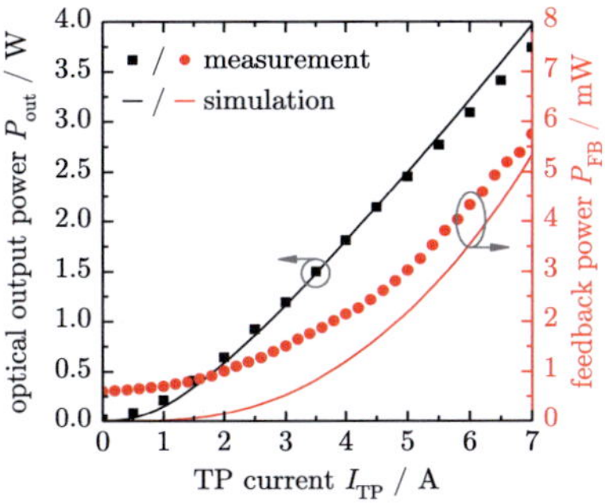

(a)

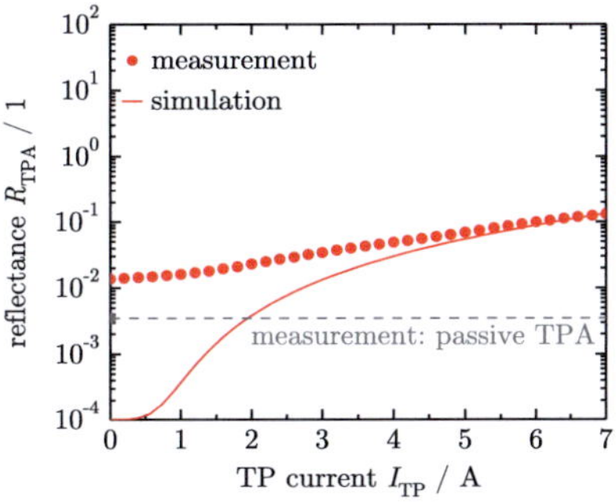

(b)

Figure 4.10.: (a) Output power and feedback power of a tapered amplifier (TPA) as a function of the taper (TP) section current at a ridge waveguide (RW) current of $I_{RW} = 600\,mA$ and a seed power of $P_{in} = 44\,mW$. (b) Corresponding reflectance of the TPA with respect to the seed power as a function of the TP current. The gray dotted line shows the reflectance of the passive TPA with no current applied to the RW and TP sections.

of the back facet. The measured reflectance at vanishing TP current is even higher than 3.5×10^{-3}, the reflectance of the passive TPA. Thus, the reflectance is not only affected by misalignment tolerances of the input beam, but probably also by spontaneous emission from the RW region.

The dependence of the output and feedback power on the RW current is shown in figure 4.11. At small currents below 150 mA the output power rises strongly with the RW current from almost 0.5 W to more than 1.2 W. Further increase of the RW current leads to an almost linear dependence of the output power, reaching 1.95 W at 900 mA. The simulation is in good agreement with the results. A completely different behavior is observed for the feedback power. Here, the feedback power stays between 2.31 mW and 1.90 mW with a slight decrease towards higher RW currents. Furthermore, the dependence of the feedback power on the RW current is characterized by tiny oscillations. The simulated feedback power shows an opposite trend, increasing with the RW current. However, oscillations are also obtained, where the local maxima, e.g. around $I_{RW} = 600\,mA$, match the measurement results. The calculated reflectance of the TPA is presented in figure 4.11b, where the measured reflectance stays around 5×10^{-2}. The simulated reflectance at zero RW current drops to about 2×10^{-3}, still more than an order of magnitude larger than the reflectivity of the back facet.

By considering the near field intensity distributions at the back facet the differences between measurement and simulation results can be explained. The insets in figure 4.11b show the respective near field intensity distributions at RW currents of 0 mA and 600 mA. The near field intensity distribution at an RW current of 600 mA is dominated by local-

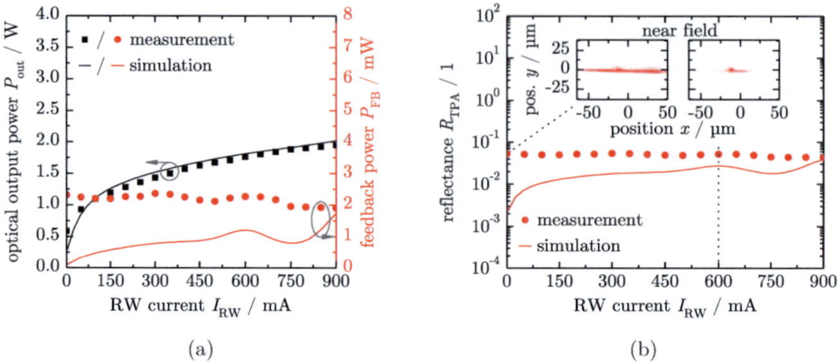

Figure 4.11.: (a) Output power and feedback power of a tapered amplifier (TPA) as a function of the ridge waveguide (RW) current at a taper (TP) current of $I_{\text{TP}} = 4\,\text{A}$ and a seed power of $P_{\text{in}} = 44\,\text{m W}$. (b) Reflectance of the TPA with respect to the seed power as a function of the RW current. The insets show the near field intensity distribution at the back facet for RW currents of $I_{\text{RW}} = 0\,\text{mA}$ and $I_{\text{RW}} = 600\,\text{mA}$.

ized emission from the RW around $x = 12\,\mu\text{m}$. Furthermore, in the horizontal direction a small side lobe near the center of the TPA ($x = 0\,\mu\text{m}$) and a weak background are visible. The near field intensity distribution at a RW current of 0 mA is very different. Here, light is emitted from a broad region in the horizontal direction within $x = \pm 50\,\mu\text{m}$. Taking into account the greatly reduced output power and hence lower feedback contributions due to reflections at the front facet at vanishing RW currents, the observed feedback emission is related to spontaneous emission from the TP section. As the RW current is increased, more optical power is emitted from the RW into the TP section, which suppresses the spontaneous emission due to better gain saturation. However, higher RW currents result in increased spontaneous emission from the RW section. This results in a balancing of the different feedback contributions and a nearly constant feedback power level.

In the following, the spatial emission characteristics are examined in more detail. In figure 4.12 the vertically integrated near and far field intensity distributions at the back facet for working points with different TP currents are shown. The near field intensity distributions, given in figure 4.12a, are characterized by an intensity maximum around $x = -11.7\,\mu\text{m}$, the center of the RW. In the measured intensity distributions, a background can be distinguished that slowly decreases to the edges of the TPA. Since the intensity level of the background is independent of the TP current, it can only be caused by spontaneous emission. As the TP current is increased from 0 A to 4 A, the additional optical feedback is emitted from a small region around the RW. Within the small region of $\pm 4\,\mu\text{m}$ around the center of the RW, the fraction of the feedback power

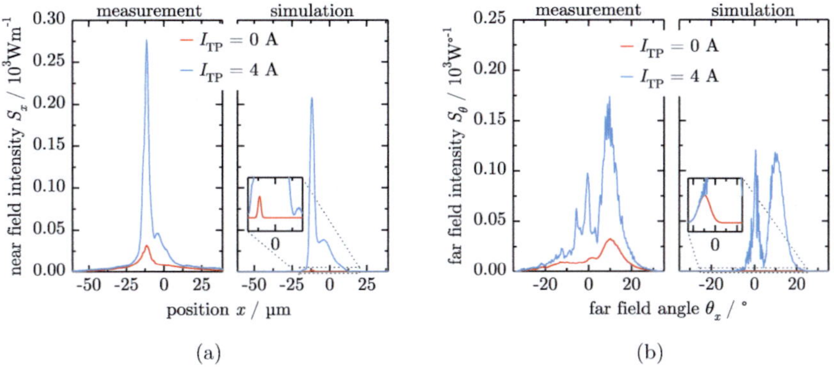

Figure 4.12.: Vertically integrated (a) near field and (b) far field intensity distributions of the feedback emission at the back facet of a tapered amplifier (TPA). The TPA is operated with a ridge waveguide current of $I_{\mathrm{RW}} = 600\,\mathrm{mA}$ and a seed power of $P_{\mathrm{in}} = 44\,\mathrm{mW}$, but two different taper currents I_{TP}.

relative to the total feedback emission rises from 0.48 to 0.62. While the near field intensity profile for zero TP current is symmetric, the distribution for a TP current of 4 A has a characteristic side lobe around $x \approx 0\,\mu\mathrm{m}$. A similar profile is obtained by the simulation, only lacking the background due to spontaneous emission. Consequently, the simulated near field intensity distribution for vanishing TP current is composed only of the reflected seed light with a Gaussian beam shape centered on the RW.

Similar aspects are also observed in the far field intensity distributions in figure 4.12b. The measured intensity distributions show a broad symmetrical background. Moreover, intensity maxima are obtained for both working points at an emission angle of 10°, which are related to the radiation of guided light in the RW. In the angular range from 5° to 15°, more than half of the feedback power is radiated, which agrees well with the power fraction within the RW region determined from the near field results. At the working point with 4 A TP current, there are additional intensity lobes around $\theta_x = 0°$, indicating emission perpendicular to the back facet. Comparable results are also obtained with the simulation, although the intensity distribution at $I_{\mathrm{TP}} = 0\,\mathrm{A}$ again differs strongly from the measurement and includes only the contribution of the reflected seed light with an emission angle of $-10°$.

To explain the near field and especially the far field intensity distributions with emission perpendicular to the back facet, the internal intensity distributions of the forward and backward propagating light are considered. In figure 4.13a the simulated vertically integrated intensity distributions at a working point with RW and TP currents of 600 mA and 4 A are given. In the intensity distribution of the forward propagating wave radiation losses of the bend of the RW are observed. Thus, due to the strong curvature of the outcoupling bend within the first 350 µm, parts of the forward propagating wave in

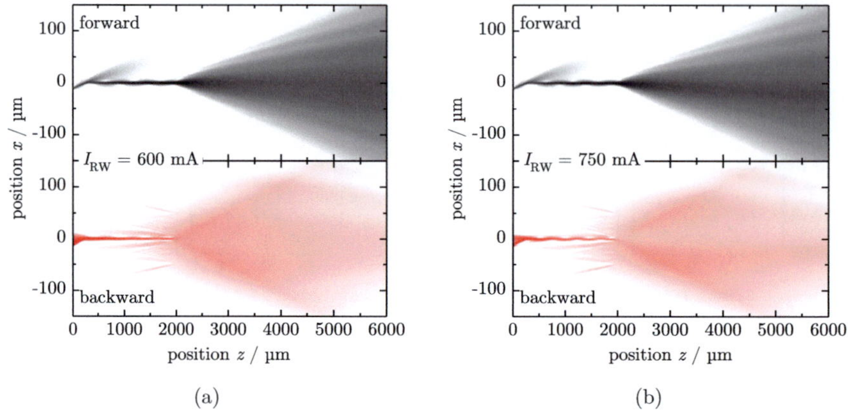

Figure 4.13.: Simulated vertically integrated intensity distributions in the horizontal-longitudinal ($x - z$) plane of the forward and backward propagating waves in a tapered amplifier (TPA). The TPA is operated with a constant seed power of $P_{\text{in}} = 44\,\text{mW}$ and a taper current of $I_{\text{TP}} = 4\,\text{A}$, but two different ridge waveguide currents I_{RW}.

the RW are radiated in the positive x-direction. Furthermore, higher-order horizontal modes are excited, causing mode beating in the RW with slightly damped oscillations toward the TP section. Despite the mode beating the light is coupled almost straight into the TP section. As a consequence, the backward propagating light is directed towards the center of the RW due to the reflection at the front facet. Finally, parts of the backward propagating light in the RW are also radiated at the bend. Therefore, the intensity components in the near field at $x = 0\,\mu\text{m}$ and in the far field at $\theta_x = 0°$ are caused by the radiation losses of the bend of the RW. Parts of the diverging backward propagating wave outside the gain regions of the TP and RW sections are strongly attenuated. Thus, significant intensity distributions at the back facet result only from the part of the backward propagating wave that is properly coupled into the RW.

The mode beating in the RW also provides an explanation for the oscillations observed in the measurement of the feedback power at varying RW currents shown in figure 4.11. In figure 4.13b the internal intensity distributions are shown for a RW current of 750 mA, which corresponds to a local minimum of the feedback power. Here, the mode beating of the forward propagating wave in the RW section leads to a slight beam steering in the TP section. The highest intensity at the front facet is obtained around $x = -25\,\mu\text{m}$. After the reflection at the front facet, the central lobe of the backward propagating wave misses the center of the RW. In addition to less power coupled to the RW, there is more intense mode beating of the backward propagating wave in the RW. Since the RW current affects the refractive index distribution in the RW, e.g. by Joule heat, the structure of the mode beating intensity pattern changes with increasing RW current, which finally causes variations in the feedback power.

4.5. Summary

The feedback emission characteristics of semiconductor optical amplifiers (SOA) under various operating conditions have been investigated in this chapter. An RWA and a TPA with design wavelengths of 1120 nm have been used as examples. At typical working points of the respective SOAs the emitted feedback power is in the lower milliwatt range resulting in reflectances relative to the seed power of the order of 1×10^{-1}. Here, the feedback emission is mainly caused by the reflection at the front facet while the overlap between the reflection and the gain region in the SOA has a significant influence on the feedback strength. The resulting feedback power increases quadratically with the optical output power, while coupling more seed power usually leads to a reduction of the feedback emission. At low optical output power of the SOA, spontaneous emission and the reflection of the seed light at the back facet can result in considerably high feedback emission. The spatial emission characteristics of the feedback are dominated by the light emission from the RW for both the RWA and the TPA. The magnitude of each effect depends on the design and operating conditions of the SOA.

In the RWA, light is guided throughout the entire gain region of the amplifier, which has only one electrode for current injection. At vanishing injection currents or optical output powers, the optical feedback is essentially due to reflection at the back facet. Due to coupling tolerances of the input seed light, a reflectance of 1×10^{-2} is obtained, which is a factor of 20 higher than the nominal reflectivity of the facet's anti-reflection coating. As the current and output power increase, the power of the optical feedback increases to 13.2 mW at typical output power levels above 1 W. Compared to the input power of 44 mW, this corresponds to a reflectance of 3.0×10^{-1}. The feedback emission increases further if the coupling of the seed power into the RW has an angular error. Even if the output power is only slightly affected, the non-optimal coupling locally leads to unsaturated optical gain, which significantly amplifies the backward propagating wave and increases the optical feedback. For a horizontal angular error of 10° relative to the ideal coupling angle, the feedback power increases more than twofold with less than 4 % loss in output power. Regardless of the coupling conditions, the optical feedback is radiated almost exclusively from the RW and at a constant angle.

The TPA has a RW section similar to the RWA, but in addition has an electrically independent TP section in which the light is not guided horizontally. With an active RW section but vanishing TP current and correspondingly low optical output power, spontaneous emission can be the dominant feedback source. With a feedback power of 0.60 mW, the corresponding reflectance of 8.4×10^{-3} is higher than both the nominal facet reflectivity and the actual reflectance of the passive TPA, which is increased by coupling tolerances. At working points with several watts of optical output power, the reflection at the front facet of the TPA contributes noticeably to the total feedback emission. While the RW current has only a small influence on the feedback and the output power, the

TP current strongly influences the optical output power and consequently the feedback power, resulting in reflectances above 1×10^{-1}. The actual feedback power of TPAs depends sensitively on the overlap of the reflection at the front facet and the gain region as well as the coupling into the RW. The corresponding backward propagating wave in the RW then also significantly determines the spatial emission behavior of the feedback, which is analogous to the feedback emission of the RWA. In contrast, the feedback part of the spontaneous emission is also radiated outside the RW.

The forward and backward propagating radiation in the SOA can be well described by a multiphysics model including a beam propagation approach. Both the associated feedback power and the spatial emission behavior can be adequately calculated. Since spontaneous emission is not included in the model, there can still be significant deviations in comparison with the experiment. For a rough estimation of only the feedback power of the SOA, a simplified model without specific material parameters can also be used.

5. Conclusions of Part I.

The feedback emission properties of periodically poled nonlinear crystals (PPNC) and semiconductor optical amplifiers (SOA) have been studied in detail in this part. In experiments and theoretical considerations, reflectances in the range of 10^{-4} to over 10^{-1} are determined in agreement. The actual reflectance depends strongly on the respective operating conditions. The feedback emission has a high spatial overlap with the incident light.

PPNCs exhibit a parasitic perturbation of the refractive index at the domain walls (DW) of adjacent domains. The DWs act as plane reflectors with an amplitude reflectivity of the order of 10^{-4} equally for both lithium niobate and lithium tantalate PPNCs. Therefore, the PPNCs act as weakly reflecting Bragg-gratings with spectrally selective reflection properties. On the one hand, there are spectrally narrow-band resonances with a high reflectance around 10^{-2}. The corresponding spectral position depends on the poling period and the PPNC temperature and has no correlation with the phase matching wavelength. On the other hand, tolerances in the poling period lead to fluctuating reflectances of the order of 10^{-4} in spectral regions far from the resonances. Considering the conversion of the incident fundamental light and the resulting depletion of the fundamental beam in the PPNC, the reflectance is expected to vary. While the reflectance for the fundamental wavelength is reduced, the generated frequency converted light is also reflected. The resonance wavelength and the phase matching wavelength have a different dependence on the PPNC properties. If applicable, the relative spectral position of the highly reflective resonances can be shifted by varying the PPNC temperature.

The optical feedback of SOAs is caused by the reflection of the incident seed light at the back and front facets and by spontaneous emission. Due to the optical gain in the SOA, the actual feedback emission and the corresponding reflectance relative to the seed power depend on the working point. The reflectance at low optical output power is of the order of 10^{-2}, where spontaneous emission can be a significant contribution to the feedback emission. The contribution due to front facet reflection increases quadratically with the output power and dominates at high power levels. On the one hand, high-power operation of SOAs can be associated with strong feedback emission, reaching reflectances of the order of 10^{-1}. On the other hand, the minimum possible feedback can be limited by spontaneous emission when the reflections at the facets are reduced, e.g. by an angled front facet. By adjusting the seed power coupled to the SOA, the feedback emission can be further reduced. While the feedback power generally decreases with increasing seed power, seed power levels below the saturation regime should be avoided as the feedback can rise by orders of magnitude.

With respect to hybrid integrated diode laser sources, the spatial emission characteristics of the different contributions to the feedback emission of PPNCs and SOAs have to be distinguished. The reflection of the incident light at the back facet is usually deflected at the angled facet. Due to the limited numerical aperture of the typical microlenses, the corresponding optical feedback does not reach the incident laser source. Under ideal coupling conditions, the other feedback contributions are directed towards the incident laser source and pass through the microlenses. In most cases the corresponding intensity distribution of the feedback has a similar shape as the incident light. Assuming a typical optical scheme, the optical feedback is expected to be coupled back into the laser source as efficiently as the initial light is coupled into the PPNC or SOA. Thus, a considerably high coupling rate to the laser source can be expected.

Considering angular coupling tolerances, the spatial emission characteristics of the feedback beam vary with respect to the incident beam. In the case of bulk PPNCs, the reflection at the plane DWs is angle-dependent and oblique incident light is reflected at an angle accordingly. Thus, the main reason for the reduction of optical feedback by tilted DWs reported in [15] is the deflection of the feedback. In contrast, the feedback emission of PPNCs or SOAs with a channel waveguide structure is determined only by the waveguide. Under oblique incidence, the direction of the feedback emission remains unchanged. Therefore, the spatial overlap between the incident beam and the feedback beam is reduced in the case of angular coupling tolerances.

Part II.

Effects of Optical Feedback

6. Tapered Amplifiers with Optical Feedback

This chapter covers the effects of varying external optical feedback on the operational behavior of tapered amplifiers (TPA). As a typical component of the high-power laser source in hybrid integrated diode laser systems, the output emission of the TPA is usually shaped by lenses and coupled to other optical elements. Periodically poled nonlinear crystals (PPNC) or optical fibers are only two examples. In chapter 3 it has been shown that PPNCs are a source of optical feedback with reflectances in the range of 10^{-5} to 10^{-2}. The typical return loss of fibers [104], [105] or residual reflectivities of antireflection coatings lead to optical feedback in a similar range. The TPA is therefore exposed to external optical feedback in typical applications.

The effect of external optical feedback on the emission behavior of TPAs depends on the characteristics of the returning feedback beam. Its properties are determined not only by the output emission characteristics of the TPA, but also by the feedback source and the optical setup used. Various feedback mechanisms and effects can be distinguished.

A common configuration of external optical feedback is the reflection of the output emission of the TPA into itself. The corresponding backward propagating wave in the TPA thus has a large spatial overlap with the forward propagating wave. Therefore, the external feedback contribution is different from the reflection at the front facet and the effect on the TPA emission behavior varies. For this configuration, a reduced optical output power and a deterioration of the beam quality of the output beam of TPAs have been observed [106], [107]. In addition, the feedback emission of the TPA itself is also increased due to external optical feedback. In contrast to these reversible effects, the device can also be permanently damaged. For broad area lasers, degradation [108] and catastrophic optical mirror damage (COMD) [109] have been identified as a result of locally increased light intensities due to external feedback.

If the output emission of the TPA is not reflected back into itself additional effects may occur. When the external optical feedback is misaligned in the horizontal direction, the resulting backward propagating wave inside the TPA does not properly overlap with the forward propagating wave. As a result, significant intensities outside the gain regions can be visible [78], [79], [88], even at the back facet. With vertical misalignment, higher-order vertical modes can be excited in the TPA, which also leads to a reduction in output power and output beam quality [110]. Even if the feedback beam is not coupled into the waveguide at all, thermal effects can still cause COMD [111].

With the typical optical scheme in hybrid integrated diode laser modules, the output emission is predominantly reflected into itself. However, the spatial beam characteristics of the output emission of the TPA, especially in the horizontal direction, depend on the working point of the TPA [107], [112]. Since the optical setup is fixed, changes in the beam shape of the output emission also alter the shape of the feedback beam returning to the TPA. Neither the dependence of the external feedback beam shape on the operating conditions of the TPA nor its effect on the output and feedback emission of the TPA have been investigated in detail yet.

The effects of external optical feedback on the operating characteristics are investigated using an exemplary TPA with a design wavelength of 1120 nm. First, the theoretical modeling of different external optical feedback beams shapes and their propagation through the TPA is described. In the second and third sections the TPA used and details of the experiment are presented. The results of the experimental and theoretical characterization of the output and feedback emission properties of the TPA are presented in the fourth section. The chapter is concluded with a summary section.

6.1. Theoretical Modeling

The theoretical modeling of tapered amplifiers (TPA) exposed to external optical feedback can be divided into two parts. First, the propagation of the output emission of the TPA to a feedback generating element and back to the TPA, including the interaction with optical elements such as lenses. Second, the interaction of the feedback light beam with the TPA.

In figure 6.1 the schematic of a typical configuration of a TPA with a source of external optical feedback is shown. The TPA is placed with the front facet at $z = l$ and emits

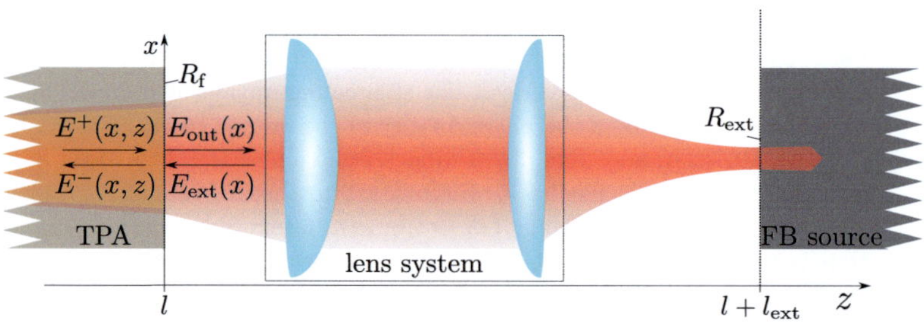

Figure 6.1.: Schematic of a tapered amplifier (TPA) in a typical optical system with a source of optical feedback (FB). $E^{\pm}$: electric field amplitude of internal propagating light, E_{out}: electric field amplitude of the output light beam, E_{ext}: electric field amplitude of optical feedback beam, R_f: front facet reflectivity, R_{ext}: external feedback reflectance, l: length of TPA, l_{ext}: distance to source of FB.

radiation that is characterized by the electric field amplitude E_{out}. The emission of the TPA propagates through a lens system and is coupled to an optical element which is also a source of optical feedback. The source of optical feedback is located at a distance l_{ext} from the front facet of the TPA and is characterized by a reflectance R_{ext}. The feedback light propagates back through the lens system and is described by the electric field amplitude E_{ext} at the front facet of the TPA. The feedback light couples to the internal propagating light with amplitudes $E^{\pm}$. The reflection of the feedback light at the front facet with residual reflectivity R_{f} and following multiple round-trips between facet and feedback source are neglected.

6.1.1. Model for External Optical Feedback

In general, the feedback beam at the front facet of the TPA can have an almost arbitrary shape, depending on the lens system and the feedback source. However, considering the typical optical configuration used in hybrid integrated diode laser systems, certain implications can be made about the feedback beam. Typically, the lens system is similar to a stretched telescope where the TPA emission is first collimated, then free space propagated and finally focused. The feedback generating optical element is placed close to the beam focus, for example to be coupled into a waveguide. Consequently, the optical feedback is emitted from the same position near the beam focus. Here, only on-axis feedback with spatial emission characteristics comparable to the incident beam, such as the feedback of periodically poled nonlinear crystals, is relevant. Off-axis feedback, such as the reflection from angled facets, does not need to be considered because the reflected light does not pass through the lens system with limited numerical aperture. Thus, the feedback generating element can be well approximated as a plane mirror in the focal plane of the focusing lens. Consequently, the output beam of the TPA is reflected into itself and the feedback beam is coupled back into the TPA without any vertical or horizontal offset.

With respect to the properties of the feedback beam and the planar geometry of the TPA, a two-dimensional theoretical model in the horizontal-longitudinal ($x-z$) plane is sufficient. The propagation of the output emission through the optical system and back to the TPA is described by the ray transfer matrix formalism. A comparable approach is commonly used for the description of laser resonators [113], [114].

In general, a ray transfer matrix with the four elements A, B, C and D describes the transformation of an paraxial optical ray passing an optical element. The displacement to the optical axis r and propagation angle u of a ray behind the optical element is related to the respective values r' and u' of the incident ray in front of the element by

$$\begin{pmatrix} r \\ u \end{pmatrix} = \begin{pmatrix} A & B \\ C & D \end{pmatrix} \cdot \begin{pmatrix} r' \\ u' \end{pmatrix} . \tag{6.1}$$

While each optical element, including free space propagation, can be modeled by an individual ray transfer matrix, their combination in an optical system can also be described by a ray transfer matrix. According to equation (6.1), the ray transfer matrix of the system is defined as the product of the individual matrices, starting the multiplication with the last optical element of the system. Thus, a single ray transfer matrix can be used to model the beam path from the TPA to the external reflector and back.

With the integral of *Collins* [115], the same ray transfer matrix can be used to calculate the transformation of arbitrary but paraxial beams based on their complex electric field distributions. In one spatial dimension the relation reads [116]

$$E(x) = e^{-\imath k_0 L_{\mathrm{opt}}} \sqrt{\frac{\imath k_0}{2\pi B}} \int E'(x') e^{-\frac{\imath k_0}{2B}\left(Ax'^2 - 2xx' + Dx^2\right)} \mathrm{d}x', \tag{6.2}$$

where E' and E are the electric field distributions in front and behind the optical system and k_0 is the vacuum wave vector. L_{opt} is the optical path length through the optical system.

To calculate the electric field amplitude E_{ext} of the feedback beam at the front facet of the TPA, equation (6.2) is adapted accordingly. With respect to the description with slowly varying amplitudes, the leading exponential phase term containing the optical path length is omitted. Furthermore, the constant external feedback reflectance R_{ext} is taken into account. Therefore, the field amplitude E_{ext} is given by

$$E_{\mathrm{ext}}(x) = \sqrt{\frac{\imath k_0 R_{\mathrm{ext}}}{2\pi B}} \int E_{\mathrm{out}}(x') e^{-\frac{\imath k_0}{2B}\left(Ax'^2 - 2xx' + Dx^2\right)} \mathrm{d}x', \tag{6.3}$$

where E_{out} is the field amplitude of the output beam at the front facet of the TPA.

6.1.2. Model for Tapered Amplifiers

The propagating light inside the TPA is described by the same multiphysics model presented in section 4.1.1. To account for the coupling to the internally propagating waves with amplitudes $E^{\pm}$ in the TPA, the boundary condition (4.3) at the front facet needs to be adapted:

$$E^{-}(x,l) = \sqrt{R_{\mathrm{f}}} E^{+}(x,l) + \sqrt{1 - R_{\mathrm{f}}} E_{\mathrm{ext}}(x). \tag{6.4}$$

The field amplitude E_{ext} depends via equation (6.3) on the field amplitude E_{out} of the output beam, which in turn is related to the field amplitude E^{+} by equation (4.4).

According to the imaging properties of the optical system, the intensity distribution at the feedback source is typically an image of the near field intensity distribution at the beam waist of the initial output beam. Therefore, the near field intensity distribution at the beam waist is used as a figure of merit for the spatial emission characteristics of the output beam. Since the horizontal beam waist of the output beam exists only virtually, located inside the TPA, the corresponding intensity distribution has to be calculated. For the calculation, a backward free-space beam propagation based on the electric field distribution E_{out} of the output emission at the front facet is used. Considering the paraxial wave equation (4.9) in free space ($\bar{n} = n = 1$), the vertically integrated near field intensity distribution S_x at the position of the horizontal beam waist z_{BW} can be calculated by the Fourier transform $\mathcal{F}$ and its inverse $\mathcal{F}^{-1}$:

$$S_x(x, z_{\mathrm{BW}}) = \frac{1}{2}\varepsilon_0 c_0 \left| \mathcal{F}^{-1}\left\{ \mathcal{F}\{E_{\mathrm{out}}\}(k_x) e^{\imath \frac{k_x^2}{2k_0} z_{\mathrm{BW}}} \right\} \right|^2. \tag{6.5}$$

The constants ε_0 and c_0 are the vacuum permittivity and the vacuum speed of light, while $k_0 = 2\pi/\lambda$ and k_x are the total vacuum and horizontal wave vectors, respectively. The horizontal beam waist position z_{BW} is relative to the position of the initial electric field distribution of the output beam. To calculate the beam waist position, the Wigner distribution function and its moments are used [116]–[118], which is described in detail in section A.7.

Estimation of Optical Feedback Power

Since the external feedback to the TPA contributes to the amplitude of the backward propagating wave, the feedback emission from the TPA can be increased. In addition to the full numerical simulation based on the multiphysics model, the total emitted feedback power of the TPA can be estimated similarly to section 4.1.2. While using the adapted boundary condition (6.4), the relation for the estimation of the feedback power reads

$$P_{\mathrm{FB}} = R_{\mathrm{b}} P_{\mathrm{in}} + (R_{\mathrm{f,eff}} + R_{\mathrm{ext,eff}}) \frac{P_{\mathrm{out}}^2}{P_{\mathrm{in}}}. \tag{6.6}$$

Again, the effective external reflectance $R_{\mathrm{ext,eff}}$ includes losses due to lack of overlap of the external feedback light with the forward propagating light in the TPA.

6.1.3. Influence of Astigmatism

The output beam of TPAs is astigmatic because the vertical beam waist of the TPA emission is at the front facet, but the horizontal beam waist is located inside the TPA. The distance between the vertical and horizontal beam waist positions is also referred to as the astigmatic difference Δz_0. Roughly estimated by $l_{\mathrm{TP}}/\bar{n}$ with the length of the taper (TP) section l_{TP} and the reference refractive index $\bar{n}$, it is usually about 1 mm. The horizontal divergence in the TP section depends on the refractive index distribution, which is dominated by thermal effects that are finally driven by the current injection [119]. Thus, the astigmatic difference varies with the working point of the TPA. While increasing the injection current within the typical tuning range, the astigmatic difference can raise by 50 µm to more than 150 µm, depending on the length of the TP section [107], [112], [119].

With respect to hybrid integrated diode laser systems with fixed optical elements, the emission of the TPA can only be collimated and properly shaped for one working point. Therefore, any deviation of the current of the TP section, e.g. for power tuning, will change the horizontal beam shape and the position of the beam waist relative to the feedback generating element. As a result, the beam is no longer reflected into itself and the beam shape of the feedback beam returning to the TPA can vary significantly. The actual variation of the feedback beam shape resulting from an altered astigmatic difference is highly dependent on the lenses used and the distance of the feedback source. However, among all reasonable configurations, three characteristic limiting cases of the feedback beam shape can be identified. These are illustrated in figure 6.2. In figure 6.2a the schematic TPA and its output emission for three different astigmatic differences is

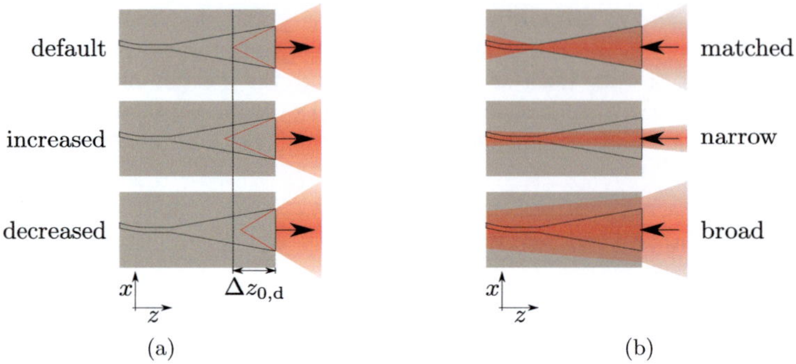

Figure 6.2.: Illustration of the influence of (a) different astigmatic differences (default, increased, decreased) of the output beam of a tapered amplifier on (b) the resulting feedback beam shape (matched, narrow, broad) in a fixed typical hybrid diode laser configuration (cf. figure 6.1). The configuration is optimized for the default astigmatic difference $\Delta z_{0,\mathrm{d}}$.

shown. The black dotted lines on the TPA symbolize the p-contact. The red solid lines show the geometric extension of the output radiation and join at the point of the virtual horizontal beam waist. With the dashed black line the position according to the default astigmatic difference $\Delta z_{0,\mathrm{d}}$ is marked as a reference. The resulting feedback beam propagating through the TPA for the three cases is visualized in figure 6.2b. For simplicity, the interactions of the feedback beam within the TPA are neglected.

At the default working point, where the output beam has the default astigmatic difference $\Delta z_{0,\mathrm{d}}$, the emission of the TPA is reflected into itself and the feedback beam shape matches the output beam shape. In this case, a large amount of the backward propagating light is coupled to the RW after converging in the TP section. The second limiting case is obtained for an increased astigmatic difference with respect to the default. Here, the feedback beam is narrow and almost collimated in the TPA. The beam diameter can be less than one third of the beam diameter of the output beam. In the third limiting case, with a decreased astigmatic difference, the feedback beam is broad and widely outshines the region below the p-contact in the entire TPA. At the front facet, the diameter of the feedback beam can be twice the beam diameter of the output beam. In the limiting cases with detuned astigmatism, only a small fraction of the backpropagating light inside the TPA is coupled back to the RW. In the following, the three feedback beam shapes are referred to as "matched", "narrow", and "broad".

6.2. Investigated Tapered Amplifier Sample

The influence of external optical feedback on the operating characteristics is exemplarily investigated by means of a TPA with a design wavelength of 1120 nm. In section 4.2, the

TPA has already been examined for feedback emission without external optical feedback and is described there in detail.

The TPA consists of a 2 mm long ridge waveguide (RW) section with a width of 4 µm, followed by a 4 mm long taper (TP) section with an opening angle of 6°. A bend in the RW section ensures that the waveguide is at an angle of 3° relative to the back facet. The antireflective coated front and back facets have residual reflectivities of 3×10^{-4} and 1×10^{-4}, respectively.

The influence of external optical feedback on the feedback emission from the TPA itself can be estimated by equation (6.6). At a typical working point of the TPA with 50 mW seed power and 2 W optical output power and matched external optical feedback with a reflectance of 10^{-3}, the feedback power is expected to be around 80 mW. For the narrow and broad feedback beam shapes with reduced effective external feedback reflectance $R_{\mathrm{ext,eff}}$, feedback power levels of a few milliwatt are reasonable.

6.3. Experimental Setup and Methods

The experimental setup for the investigation of external optical feedback effects on TPAs is an extension of the setup shown in section 4.3 in figure 4.5. All additional components are placed in front of the TPA and only affect the output emission. A schematic of the setup is given in figure 6.3. The TPA is mounted on the heat sink with an actively

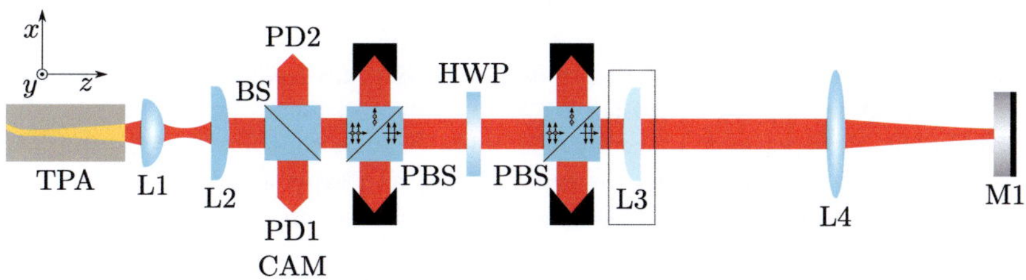

Figure 6.3.: Schematic of the experimental setup for the characterization of tapered amplifiers (TPA) with defined external optical feedback. L1: aspherical lens, L2: cylindrical lens, BS: beamsplitter, PBS: polarizing beam splitter, HWP: half wave plate, L3: optional cylindrical lens, L4: spherical lens, M1: plane mirror, PD1 / PD2: power detectors, CAM: CCD camera.

controlled temperature of 25 °C. The output emission of the TPA is collimated in the vertical (y) direction by an aspherical lens L1 with a focal length of 8 mm. Since the output beam of the TPA is astigmatic, a second cylindrical lens L2 with a focal length of 71 mm is used to collimate the beam in the horizontal (x) direction. The collimated beam is partly reflected by a beamsplitter (BS) with a reflectivity of $R_{\mathrm{BS}} = 0.44$ to characterize the output emission of the TPA. While the optical power is measured by the power detector PD1 (*UP19K-15S-W5*) a CCD camera (CAM) *Scorpion SCOR-20-SOM* is used to record the near field intensity distribution. The optical elements used for the imaging are not shown here, but are explained in section C.4. The transmitted beam through the beamsplitter BS passes a combination of two polarizing beamsplitters (PBS) and a half wave plate (HWP), while the reflections at the PBS are directed to beam dumps. Behind the latter PBS, an optional cylindrical lens L3 with a focal length of 500 mm or 250 mm can be placed to introduce additional beam shaping in the horizontal direction. With a spherical lens L4 with a focal length of 250 mm and a plane silver mirror M1, the beam is reflected into itself. The mirror M1 is placed near the focal plane of L4, resulting in a total distance to the front facet of the TPA of about $l_{\mathrm{ext}} = 960\,\mathrm{mm}$. By the combination of the PBS and HWP the beam is attenuated again and is partially reflected from the BS to the power detector PD2 (*PH20-Ge-OD2*). The transmitted part of the beam is coupled back to the TPA by L1 and L2.

By varying the orientation of the HWP, the passing beam can be attenuated and a defined power of the optical feedback coupled to the TPA and corresponding external feedback reflectance can be set. The external feedback reflectance R_{ext} is defined as the ratio of the optical power P_{ext} of the external feedback to the output power P_{out} of the TPA. With P_{PD1}, the measured power from power detector PD1, and the reflectivity R_{BS} of the beamsplitter BS, the output power of the TPA is given by $P_{\mathrm{out}} = P_{\mathrm{PD1}}/R_{\mathrm{BS}}$. The power P_{ext} of the external feedback is given by $P_{\mathrm{ext}} = (1 - R_{\mathrm{BS}})P_{\mathrm{PD2}}/R_{\mathrm{BS}}$, where P_{PD2} is the measured power from power detector PD2. Consequently, the relation for the external feedback reflectance reads

$$R_{\mathrm{ext}} = (1 - R_{\mathrm{BS}})\frac{P_{\mathrm{PD2}}}{P_{\mathrm{PD1}}}. \tag{6.7}$$

With respect to the optical components used, the external feedback reflectance can be tuned in a range from 1×10^{-6} to 1×10^{-2}.

Throughout all experiments, the TPA is operated at a constant working point with a seed power of 44 mW and ridge waveguide (RW) and taper (TP) section currents of 600 mA and 4 A, respectively. At that working point, the output power of the solitary TPA is 1.8 W.

To obtain a defined external optical feedback condition in terms of reflectance and beam shape, the setup must be adjusted accordingly. In the default feedback configuration without lens L3, the output beam from the TPA is reflected back into itself, resulting in the matched beam shape. First, the emitted light from the TPA is collimated at the desired working point in the absence of external optical feedback. Second, the optical feedback is set to the highest possible external feedback reflectance of $R_{\mathrm{ext}} = 1 \times 10^{-2}$. While measuring the feedback emission of the TPA (described in section 4.3), the position and angle of the mirror M1 are adjusted. The highest TPA feedback power is obtained when the coupling of the feedback beam to the RW is maximized, which corresponds to the desired matched feedback beam shape. The configuration with lens L4 and mirror M1 not only ensures the matched feedback beam shape in the horizontal direction, but also achieves a reflection into itself in the vertical direction. According to the calculation by the optical simulation software *BeamXpert Designer*, which includes a ray tracing method, a coupling efficiency of more than 0.99 can be achieved. Even taking into account coupling losses due to misalignment and tolerances of the optical elements are concerned, the coupling efficiency can be considered to be higher than 0.7. Compared to the total tuning range of the external feedback reflectance, the influence of tolerances can be neglected.

In order to realize feedback configurations resulting in the narrow or broad feedback beam shapes, the setup is first adjusted similar to the matched feedback beam shape configuration. In a second step, the lens L3 is inserted and adjusted to maximize the TPA feedback emission. For the narrow feedback beam shape, L3 with a focal length of 500 mm is used. According to the calculations with *BeamXpert Designer*, the corresponding feedback beam shape has a beam diameter of 198 µm at the front facet, which remains almost unchanged within the TPA. The broad feedback beam shape is obtained using L3 with a focal length of 250 mm. Here, the broad feedback beam converges in the TPA from 816 µm diameter at the front facet to 313 µm at the back facet. Since the feedback beam shape is adjusted independently of the collimation of the output beam, the measurement of the emission characteristics of the TPA is not disturbed. In addition, the TPA can be operated at the same working point for all measurements, allowing direct comparability of the results.

At each external optical feedback configuration, the output and feedback emission characteristics of the TPA are determined. The output emission properties are characterized with respect to the optical output power P_{out} and the near field intensity distribution at the horizontal beam waist. Here, the imaging by lens L1 and the resulting horizontal intermediate focus must be considered (see section C.4). With the astigmatic difference $\Delta z_0 = 1.33\,\mathrm{mm}$ of the output beam and the focal length $f_{\mathrm{L1}} = 8\,\mathrm{mm}$ of the lens L1, an additional factor of 6.02 has to be taken into account for the magnification of the beam waist intensity distribution.

The methods for the characterization of the optical feedback of the TPA are described in detail in section 4.3. The used material parameters for the simulation with the multiphysics model are given in tables B.3 and B.5. In table B.6 the ray transfer matrix elements A, B, C and D for the three external feedback configurations are provided. The ray transfer matrix elements are calculated for the configurations of the experimental setup using the *BeamXpert Designer* software.

6.4. Results

First, the output emission behavior of the tapered amplifier (TPA) under external optical feedback with varying external feedback reflectance and feedback beam shapes is discussed. The optical output power P_{out} of the TPA is evaluated as the relative optical power $P_{\mathrm{rel}} = P_{\mathrm{out}}/P_{\mathrm{out,\ w/o\ FB}}$ with respect to the optical output power $P_{\mathrm{out,\ w/o\ FB}}$ of the solitary TPA without external optical feedback. In figure 6.4a the relative optical output power of the TPA as a function of the external feedback reflectance for the three feedback beam shapes is shown.

The optical output power remains constant within the measurement uncertainty up to an external feedback reflectance of 1×10^{-4}. As the external feedback reflectance increases, the optical output power begins to decrease, with a significant decrease oc-

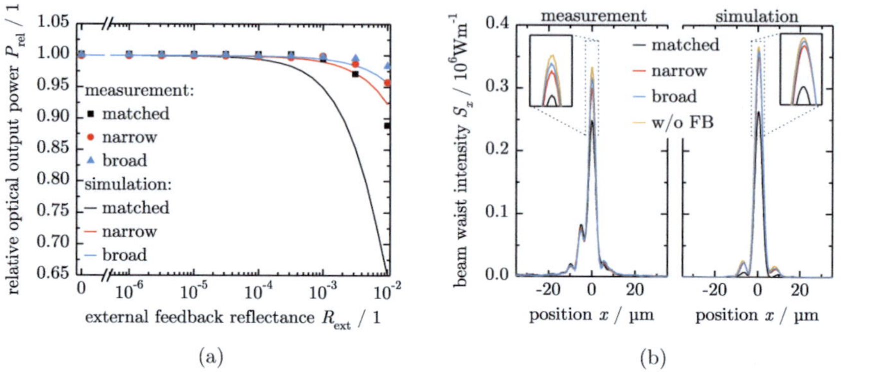

Figure 6.4.: (a) Relative optical output power of the tapered amplifier (TPA) under external feedback with three beam shapes as a function of the external feedback reflectance. The relative optical power is defined as the ratio of the output power P_{out} to the output power $P_{\text{out, w/o FB}}$ of the solitary TPA without external optical feedback. (b) Vertically integrated near field intensity distribution at the beam waist of the output emission of the TPA for three external feedback beam shapes with an external feedback reflectance of $R_{\text{ext}} = 1 \times 10^{-2}$ and no external feedback (FB). The TPA is operated at a constant working point with a seed power of 44 mW and ridge waveguide and taper section currents of 600 mA and 4 A.

curring only for external feedback reflectances greater than 1×10^{-3}. At the highest external feedback reflectance of 1×10^{-2}, the optical output power decreases by 11.1 % in the case of the matched beam shape. The narrow and broad feedback beam shapes cause a much smaller reduction in optical output power of 4.3 % and 1.7 %, respectively. The simulation results show the same qualitative behavior with the largest power reduction for the matched beam shape and only a small power reduction for the broad beam shape. However, the relative optical output power is overestimated by up to a factor of three.

The influence of the external optical feedback on the vertically integrated near field intensity distribution at the beam waist of the output beam of the TPA is shown in figure 6.4b. In addition to the feedback configurations with the maximum external feedback reflectance of 1×10^{-2}, the near field intensity distribution without external feedback is shown for comparison. The measured near field intensity distributions are characterized by a central lobe and a significant side lobe around a horizontal position of $x \approx -5\,\mu\text{m}$. With external optical feedback, the intensity of the central lobe is reduced and shows the same dependence on the feedback beam shape as the total power loss in figure 6.4a. In contrast, the side lobes of the near field intensity distribution remain almost unchanged. The decrease of the power content in the central lobe can be associated with a deterioration of the spatial beam quality. The simulated near field intensity distribution also consists of an intense central lobe and slightly asymmetrically distributed side lobes. Under the influence of external optical feedback, the central intensity lobe shows almost the same decrease as in the measured data. However, the side lobes scale differently because the corresponding intensity also decreases, at least for the matched beam shape. Therefore, the external optical feedback mainly affects the central lobe of the near field intensity distribution, while the general shape remains almost unchanged. Considering only the power loss in the central intensity lobe within $x = -4\,\mu m$ and $x = 4\,\mu m$, a much better agreement between measurement and simulation is obtained. At the maximum external feedback reflectance, the measured relative optical power in the central lobe decreases by 22 %, 9 %, and 4 % for matched, narrow, and broad feedback beam shapes. The simulation gives a reduction of 31 %, 7 %, and 3 %, respectively.

Next, the feedback emission characteristics of the TPA under the influence of external optical feedback are investigated. The total optical feedback power emitted from the TPA and the corresponding TPA reflectance as a function of the external optical feedback reflectance are given in figure 6.5. The emitted feedback power is much more sensitive

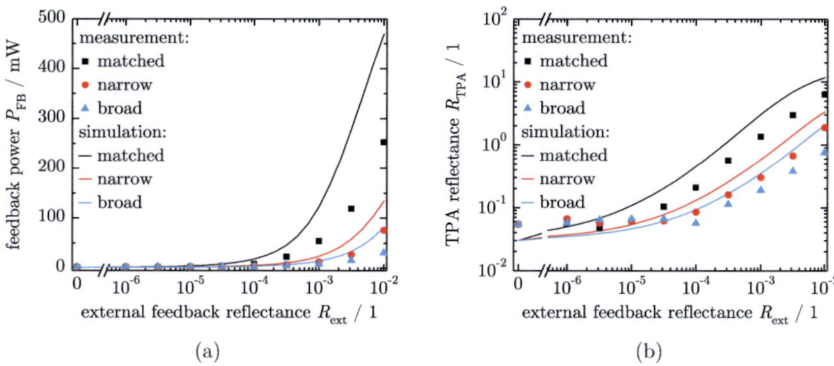

Figure 6.5.: (a) Feedback power and (b) the corresponding reflectance with respect to the seed power of the tapered amplifier (TPA) exposed to external feedback with three beam shapes as a function of the external feedback reflectance. The TPA is operated at a constant working point with a seed power of 44 mW and ridge waveguide and taper section currents of 600 mA and 4 A.

to external feedback than the respective optical output power of the TPA. Here, a doubling of the feedback power for external feedback reflectances of 3×10^{-5} is already visible for the matched feedback beam shape. Especially in the logarithmic scale for the TPA reflectance in figure 6.5b the increasing TPA reflectance is characterized by a clear threshold. Similar to the output power, the strongest and weakest influence is found for the matched and broad beam shapes, respectively. From an initial value of 2.16 mW, the feedback power raises almost exponentially by orders of magnitude to a maximum between 30.2 mW and 252.6 mW. Thus, the associated TPA reflectance actually exceeds unity and approaches 6. Both the optical feedback power of the TPA and its dependence on the external feedback beam shape suit the estimations by the simplified model.

The simulation again leads to qualitatively the same results, while the experimental results are overestimated by up to a factor of two. In contrast to the measured data, the TPA reflectance increases slightly already at low external feedback reflectances and the transition to a stronger increase is smoother.

The feedback emission for vanishing or low external optical feedback has a high contribution of spontaneous emission (cf. section 4.4.2). Thus, the influence of the external optical feedback becomes apparent only when the resulting additional feedback emission of the TPA exceeds the power fraction of the spontaneous emission. In this case, the feedback emitted from the TPA is strongly dominated by the external feedback light, and the contributions from the facets and spontaneous emission become negligible.

The correlation between the reduction of the optical output power and the increase of the feedback emission is further investigated based on the power of the forward and backward propagating waves in the TPA. The simulated optical power of the forward propagating wave as a function of the longitudinal position in the TPA for different feedback beam shapes and external feedback reflectances is shown in figure 6.6a. Except

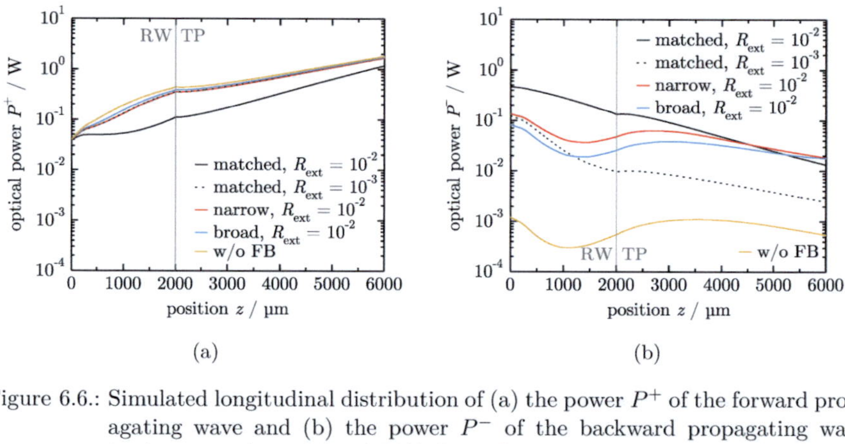

Figure 6.6.: Simulated longitudinal distribution of (a) the power P^+ of the forward propagating wave and (b) the power P^- of the backward propagating wave in the tapered amplifier (TPA) for different external feedback (FB) beam shapes and external feedback reflectances R_{ext}. The solid gray line marks the boundary between the ridge waveguide (RW) and the taper (TP) sections. The TPA is operated at a constant working point with a seed power of 44 mW and RW and TP section currents of 600 mA and 4 A.

for the matched beam shape and an external feedback reflectance of $R_{\text{ext}} = 1 \times 10^{-2}$, the optical power P^+ of the forward propagating wave increases by one order of magnitude in the ridge waveguide (RW) section. The individual trends are slightly different, with the highest optical power obtained without external optical feedback. In the taper (TP) section, the optical power increases further by about 1.3 W, while no significant differences can be observed. In the case of the matched feedback beam shape and an external feedback reflectance of 1×10^{-2}, a different behavior can be distinguished. In the RW section, the optical power remains almost constant in the first half and increases only by 50 mW in the second half. Finally, the optical power in the TP section is increased by more than 1 W, with almost the same longitudinal trend as in the previous cases. Thus, the large decrease in the optical output power of the TPA is caused by the strongly reduced amplification in the RW section. The effect varies strongly with the external feedback reflectance. For the matched beam shape and an external feedback reflectance of 1×10^{-3}, there are only small changes in the power evolution in the RW section compared to the solitary TPA without external feedback.

In figure 6.6b the optical power of the backward propagating wave as a function of the longitudinal position is shown. Unlike the power of the forward propagating waves, the power of the backward propagating waves varies greatly for each feedback configuration. Without external optical feedback, the optical power shows a variable pattern due to the changing overlap of the backward propagating wave and the gain section. The corresponding vertically integrated intensity distribution of the backward propagating wave in the horizontal-longitudinal ($x - z$) plane is given in figure 6.7a for reference (cf. figure 4.13a in section 4.4.2). Near the front facet, a large fraction of the wave still propagates in the gain region, resulting in an overall increase in power. Due to divergence and the narrowing of the gain region toward the RW section, the overlap decreases and absorption dominates. Consequently, the power decreases. With only a small fraction coupled into the RW, the power increase in the waveguide is initially small, while light outside the gain region continues to be strongly attenuated. Near the back facet, the power of the backward propagating wave is mainly confined in the RW and the power increases due to the optical gain.

With external optical feedback with the narrow or broad beam shape, the corresponding optical power has almost the same characteristic as without external feedback. The vertically integrated intensity distributions of the backward propagating waves are given in figure 6.7b. In contrast to the broad feedback beam shape, the narrow beam shape results in a backward propagating wave that is in the gain region up to about the center of the TP section ($z = 4000\,\mu\text{m}$). Thus, the associated optical power first increases more and then decreases closer to the RW section.

In the case of the matched feedback beam shape, a different behavior is obtained for the power of the backward propagating wave. Since the backward propagating wave has an ideal overlap with the gain region, the optical power increases almost monotonically in the TPA for both external feedback reflectances of 1×10^{-3} and 1×10^{-2}. However, there are clear differences in the RW section. At an external feedback reflectance of 1×10^{-3}, the power increases strongly by an order of magnitude, but reaches a plateau near the back facet, which occurs similarly for the narrow and the broad feedback beam shapes. For the high external feedback reflectance of 1×10^{-2}, the power increase in

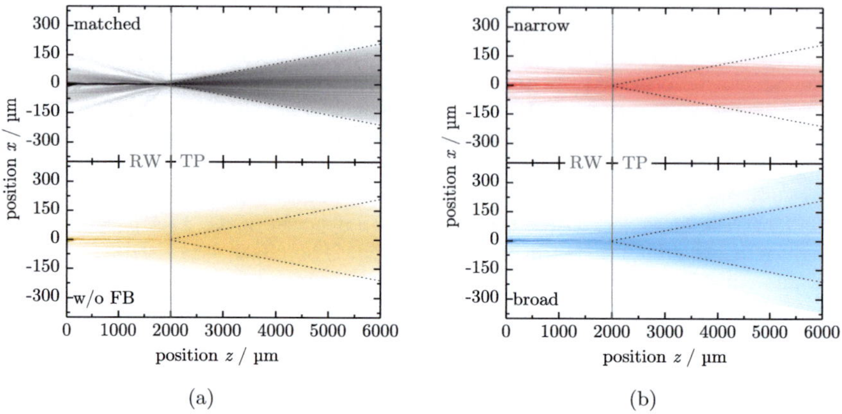

Figure 6.7.: Simulated vertically integrated intensity distributions in the horizontal-longitudinal ($x - z$) plane of the backward propagating waves in the tapered amplifier (TPA) for different external feedback beam shapes with an external reflectance of $R_{\text{ext}} = 1 \times 10^{-2}$ and no external feedback (FB). The solid gray line marks the boundary between the ridge waveguide (RW) and the taper (TP) sections, while the dotted lines indicate the gain region in the TP section. The TPA is operated at a constant working point with a seed power of 44 mW and RW and TP section currents of 600 mA and 4 A.

the RW section is much smaller. Thus, the optical gain is strongly saturated by the high power of the backward propagating wave and reduces the amplification of the much weaker forward propagating wave.

The intensities of the backward propagating waves shown in figure 6.7 indicate that small contributions far outside the RW reach to the back facet even for the matched feedback beam shape. Therefore, the effects of the external feedback on the spatial emission characteristics of the feedback of the TPA are examined below. Figure 6.8a shows the vertically integrated near field intensity distributions at the back facet of the TPA for vanishing and high external optical feedback with three feedback beam shapes and an external feedback reflectance of 1×10^{-2}. For a better comparison, the intensity distributions are also scaled to their maximum and given in figure 6.8b.

In all cases, the near field intensity distributions consist of an intense lobe at the position of the RW around $x = -11.7\,\mu\text{m}$ and significant side lobes. While the relative height with respect to the intensity maximum is almost constant for all feedback configurations, different shapes of the side lobes can be distinguished. For the narrow and broad feedback beam shapes, the side lobes have a substructure with multiple local maxima extending even to positions around $x = 20\,\mu\text{m}$. Furthermore, weak intensity lobes are visible at $x = \pm 40\,\mu\text{m}$ for all types of external optical feedback. The appearance of these

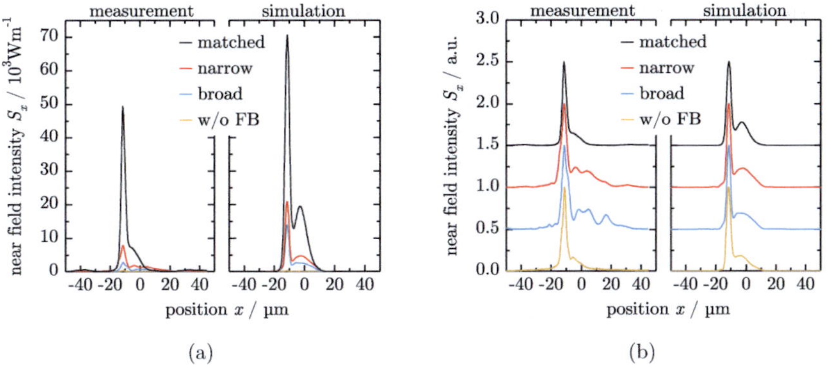

Figure 6.8.: Vertically integrated near field intensity distributions of the feedback emission at the back facet of the tapered amplifier (TPA) for different external feedback beam shapes with an external reflectance of $R_{\text{ext}} = 1 \times 10^{-2}$ and no external feedback (FB). The intensities are normalized (a) to the total feedback power and (b) to their respective maximum intensity. The TPA is operated at a constant working point with a seed power of 44 mW and ridge waveguide and taper section currents of 600 mA and 4 A.

small lobes is consistent with the simulated two-dimensional intensity distributions in figure 6.7. Since the feedback emission is dominated by the external optical feedback, a smooth background due to spontaneous emission is not visible in the corresponding near field intensity distributions.

With the observed differences in the side lobes not only the shape of the intensity distributions varies, but also the relative power in the central lobe. The power fraction within ±4 µm around the center of the RW is 0.68 for the matched feedback beam shape, which is comparable to the solitary TPA without external feedback. For the narrow and broad feedback beam shapes, the power fraction decreases slightly to 0.55 and 0.51, respectively, due to the side lobes.

The numerical simulation gives comparable near field intensity distributions with an intense maximum and characteristic side lobes. However, much less difference is observed for the varying external feedback conditions. With respect to the maximum intensity, side lobes far away from the center are rarely visible. Although the attenuation of the backward propagating wave outside the RW section is overestimated by the theoretical model, the power fraction in the central lobe agrees with the measurements. Here, the power fractions are 0.63 and 0.60 for vanishing and matched external feedback, while lower ratios of 0.53 and 0.56 are obtained for narrow and broad feedback beam shapes.

Although the external optical feedback affects the near field intensity distribution of the feedback at the back facet of the TPA, it is still dominated by the emission of the light guided in the RW. Again, the most intense lobe is caused by the direct emission from the RW, while the side lobe is mainly caused by the radiation losses in the bend. A consistent result is obtained by considering the corresponding vertically integrated far field intensity distributions given in figure 6.9a. Similar to the near field intensity

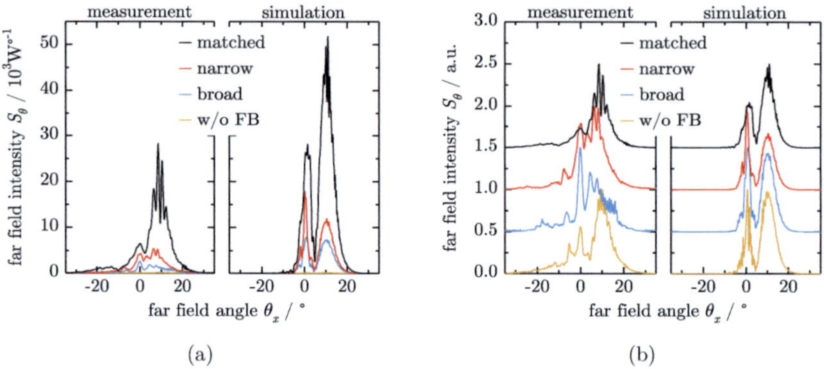

Figure 6.9.: Vertically integrated far field intensity distributions of the feedback emission at the back facet of the tapered amplifier (TPA) for different external feedback beam shapes with an external reflectance of $R_{\text{ext}} = 1 \times 10^{-2}$ and no external feedback (FB). The intensities are normalized (a) to the total feedback power and (b) to their respective maximum. The TPA is operated at a constant working point with a seed power of 44 mW and ridge waveguide and taper section currents of 600 mA and 4 A.

distributions, the far field intensity is also scaled to the respective maximum, as shown in figure 6.9b.

Corresponding to the emission of guided light in the RW, again a large part of the feedback is emitted at angles around 10°. In contrast, the radiation losses of the bent waveguide, but also the external optical feedback, lead to the central intensity distribution around 0°. In the far field intensity distributions, strong modulations can be found, so that the relative height of the local maxima can differ strongly. The same behavior can be seen in the simulation results. However, despite the strongly varying intensity patterns, the power fraction in the range between 5° and 15° changes only slightly for the different feedback configurations. In the measurement and simulation, the power fractions determined from the far field intensity distributions agree with the corresponding power fractions from the near field intensity distributions around the RW position. Thus, there are no fundamental differences in the effects due to different feedback beam shapes. The main influences on the output and feedback emission of the TPA are still determined by the coupling efficiency of the feedback beam into the RW.

6.5. Summary

In this chapter, the influence of external optical feedback on the emission behavior of a tapered amplifier (TPA) with an emission wavelength around 1120 nm has been investigated. Both the output emission and the feedback emission from the TPA have been characterized under external feedback with different beam shapes up to an external feedback reflectance of 1×10^{-2}. The strength of the influence depends mainly on the coupling of the external feedback beam into the ridge waveguide (RW). Therefore, the maximum influence is found for external feedback that is generated by a back reflection of the output emission into itself. At the maximum external feedback reflectance, the optical output power is reduced by just over 10 %, while the feedback emission from the TPA rises to power levels as high as six times the input seed power. With the exception of small effects, the spatial emission behavior of the output and feedback emission from the TPA is not significantly changed by the external feedback. In particular, the emission behavior of the feedback of the TPA is always dominated by the emission of the light guided in the RW.

The external optical feedback reaching the TPA depends not only on the feedback source, but also on the emission characteristics of the TPA itself. Especially in a fixed optical setup, the working point dependent astigmatism of the output emission of the TPA can cause the beam shape of the external feedback to vary in the horizontal direction. Three scenarios can be distinguished with matched, narrow, and broad feedback beam shapes. The standard case is the matched beam shape, where the output emission of the TPA is reflected into itself. As the astigmatism changes, different external feedback configurations result with a narrow collimated or a broad overshining horizontal shape of the external feedback beam at the TPA.

The output characteristics of the TPA are only slightly dependent on the external feedback reflectance. While the output power is always reduced with external optical feedback, the reduction is negligible for external feedback reflectances up to 1×10^{-3}. The power decrease is mainly due to the gain saturation in the RW section by the feedback-induced backward propagating wave in the TPA. Therefore, for different feedback beam shapes, only the coupling of the backward propagating wave into the RW is critical for influencing the output power. With an external feedback reflectance of 1×10^{-2} and almost complete coupling in the RW (matched beam shape), the output power decreases by 11.1 %. The power decrease is consequently smaller for other feedback beam shapes. The beam shape of the output emission does not change fundamentally. Considering the near field intensity distribution at the beam waist of the output beam, only the relative power contained in the central lobe decreases, which mainly corresponds to the decrease of the optical output power.

The feedback emission from the TPA itself is very sensitive to external optical feedback. Already at external optical feedback reflectances of more than 1×10^{-5}, the feedback power can increase by a factor of two and continues to increase almost exponentially. Again, the coupling of the external feedback into the RW has a significant effect on the feedback power. At an external feedback reflectance of 1×10^{-2}, the feedback power reaches values between 30.2 mW and 252.6 mW for the broad and the matched beam shapes, respectively. This corresponds to reflectances of the TPA from 6.8×10^{-1} to 5.7. Once the main contribution to the feedback emission of the TPA is dominated by the external optical feedback, the emission of spontaneous emission is negligible.

The spatial emission pattern of the optical feedback of the TPA is determined by the emission pattern of the RW for all external feedback configurations. Thus, the feedback emission from the TPA is still directed to the incident seed power source. However, there are qualitative and quantitative differences for different feedback beam shapes. Since only a part of the backward propagating wave in the TPA is coupled into the RW for the narrow and broad beam shapes, there are also weak intensity distributions outside the RW. As a result, the intensity distributions can be strongly modulated and the actual fraction of the feedback power emitted from the RW region can be reduced by about 10 percentage points.

With the numerical simulation based on a multiphysics model, the experimentally observed behavior is qualitatively well described in all aspects. The effects on the output power and the feedback power of the TPA are quantitatively overestimated, but the simulation provides important insights for understanding the individual feedback effects. If only the effect of the external feedback on the feedback emission of the TPA is of interest, the simplified model based on directly accessible quantities is still sufficient.

7. Distributed Bragg-Reflector Ridge Waveguide Lasers with Optical Feedback

In this chapter, the effects of external optical feedback on the emission characteristics of distributed Bragg-reflector ridge waveguide lasers (DBR-RWL) are investigated. DBR-RWLs are edge-emitting diode lasers that emit a single spectral and spatial mode with up to watt-level optical output power. By optimizing the laser cavity geometry accordingly, a wide range from several milliwatts to several hundred milliwatts of optical output power can be covered. In typical application scenarios, DBR-RWLs are used both as a direct high-power laser source and as a seed laser source for master oscillator power amplifier systems. Thus, different sources of optical feedback are possible. Examples are optical fibers [104], [105], periodically poled nonlinear crystals (see chapter 3), or semiconductor optical amplifiers (SOA, see chapters 4 and 6), which are characterized by external reflectances of 1×10^{-5} to 1×10^{-1} and even higher. Although optical isolators can be used to attenuate the feedback, their use in hybrid integrated miniaturized laser systems is limited for low optical power and specific emission wavelength ranges.

Single spectral mode emitting diode lasers exposed to external optical feedback have been studied experimentally and theoretically in numerous works. Among different types of diode lasers such as Fabry-Pérot lasers [120], distributed feedback lasers [121] or even high-power distributed Bragg-reflector tapered diode lasers [122], comparable general emission behavior under external feedback is found [123], [124].

With weak external optical feedback, the emission characteristics of the diode laser are slightly perturbed, depending on the actual feedback conditions. The optical output power, the emission wavelength, or the spectral width of the emission are thereby affected in a strongly deterministic manner. The variations are in good agreement with theoretical predictions [122], [125] and can even be used for precise measurements [126].

As the external feedback strength increases, the emission behavior of the laser becomes more complex. While initially many dynamic effects occur, a completely chaotic emission state, also known as coherence collapse, can also be reached [127]–[129]. The nonlinear dynamics and chaotic states are still in agreement with theoretical considerations, but much more numerical and theoretical effort is required. In order to properly characterize the emission behavior at higher feedback strengths, a detailed experimental and theoretical analysis is therefore necessary in each individual case.

The external optical feedback is usually assumed to be coherent with the laser emission. With respect to the spontaneous emission from SOAs, incoherent external feedback is also possible. Although the effects on the emission behavior may well vary, incoherent optical feedback also leads to variations in the emission behavior and dynamic effects up to chaotic behavior. However, the critical external feedback strengths are typically higher compared to coherent optical feedback [130]–[132].

The influence of external optical feedback on the emission behavior of DBR-RWLs is investigated using DBR-RWLs with an emission wavelength of 1120 nm as an example. Under coherent external optical feedback up to reflectances of 1.5×10^{-1}, different cavity configurations of the DBR-RWLs are compared. The first section presents the theoretical modeling based on the well-known Lang-Kobayashi equations and their adaptation to DBR-RWLs. In the second section, the investigated DBR-RWLs, which are characterized by different resonator geometries are presented. The experimental setup is described in the third section, while the experimental and theoretical results are presented in the fourth section. The results of this chapter are summarized in the last section.

7.1. Theoretical Modeling

In distributed Bragg-reflector ridge waveguide lasers (DBR-RWL), the light is guided by a ridge waveguide in the entire device. Thus, in addition to horizontal waveguiding, vertical waveguiding is realized and only a single spatial mode is emitted, which is usually independent of the actual working point. In most applications, the emission from DBR-RWLs is shaped by lenses and coupled into other optical components such as semiconductor optical amplifiers or periodically poled nonlinear crystals. As described earlier when considering feedback on tapered amplifiers in section 6.1, in such configurations the output emission is reflected back into itself. Thus, DBR-RWLs with external optical feedback can be described with an effective one-dimensional model.

In figure 7.1 the configuration of a DBR-RWL and an external optical feedback source is schematically shown. The DBR-RWL consists of a distributed Bragg-reflector (DBR)

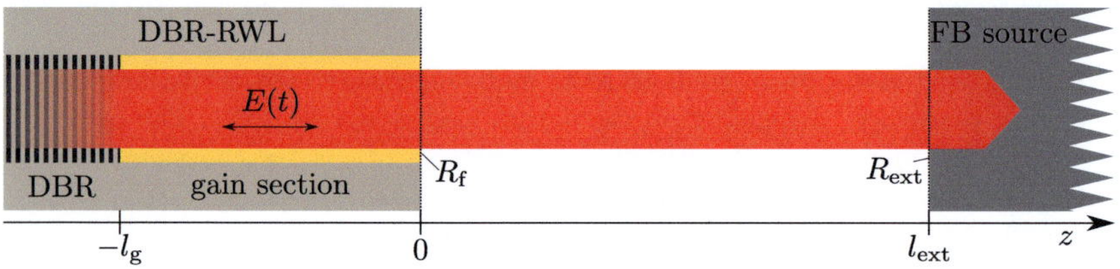

Figure 7.1.: Schematic configuration of a distributed Bragg-reflector ridge waveguide laser (DBR-RWL) and a source of external optical feedback (FB). DBR: distributed Bragg-reflector, l_g: gain section length, l_{ext}: distance to FB source, R_f: front facet reflectivity, R_{ext}: external feedback reflectance, E: electric field amplitude.

section and a gain section with a length of l_g. The light inside the DBR-RWL is described by the complex electric field amplitude E, while the output emission is coupled out through the front facet with reflectivity R_f at $z = 0$. At a distance l_{ext} from the front facet, the emission from the DBR-RWL is partially reflected by the feedback (FB) source with a reflectance of R_{ext}.

In solitary laser operation, the electric field inside the internal laser cavity, formed by the DBR and the front facet, is the solution to a resonance condition, yielding the well-known laser conditions. Thereby the emission characteristics such as the output power and the emission wavelength are defined. With an external reflector, an external cavity is formed, competing with the internal laser cavity. The laser conditions of the compound cavity result in different solutions for the electric field.

There are several approaches to mathematical modeling that differ in the description of the coupling of the external optical feedback to the field in the internal laser cavity. Assuming that the external optical feedback is a small perturbation to the field in the internal laser cavity, the well-known Lang-Kobayashi equations [120] can be derived. Alternative models [133], [134] can be derived from the theory of mode locking [135], [136] if the external reflector is considered as a part of the laser cavity. Although the Lang-Kobayashi equations are valid only for small external feedback reflectances, they allow for a simple description and have become one of the most common approaches to analyze optical feedback phenomena [125], [137], [138].

7.1.1. Lang-Kobayashi Equations

The influence of external optical feedback on the emission behavior of single spectral mode emitting diode lasers can be described by a set of delay-differential equations [120] for the electric field amplitude E and the carrier density N:

$$\partial_t E(t) = \left[\imath\omega_\mathrm{l} + (1+\imath\alpha_\mathrm{H})\frac{c_0}{2n_\mathrm{g}}(g(N(t)) - g_\mathrm{l}) \right] E(t) + \eta_\mathrm{ext} E(t-\tau_\mathrm{ext}) \tag{7.1}$$

$$\partial_t N(t) = \frac{j}{e_0 d_\mathrm{AZ}} - \frac{N(t)}{\tau_N} - \frac{g(N(t)) c_0 |E(t)|^2}{n_\mathrm{g}}. \tag{7.2}$$

ω_l and g_l are the oscillation frequency and the threshold gain of the lasing mode of the solitary laser. α_H and n_g are the material-dependent linewidth enhancement factor and the group refractive index, while c_0 is the vacuum speed of light. The external optical feedback is coupled to the internal laser cavity with the coupling rate η_ext and a time delay of the external round-trip time $\tau_\mathrm{ext} = 2l_\mathrm{ext}/c_0$. The threshold gain g of the lasing mode with external feedback is defined by the carrier density obeying the rate equation (7.2) (cf. equation (4.15) in section 4.1). j is the injection current density, e_0 the elementary charge, and d_AZ is the thickness of the active zone. Spontaneous recombination is modeled by the carrier lifetime τ_N. The electric field is normalized so that $|E|^2$ is a photon density.

The coupling rate of the external feedback is given by

$$\eta_\mathrm{ext} = \frac{1-R_\mathrm{f}}{\tau_\mathrm{in}}\sqrt{\frac{R_\mathrm{ext}}{R_\mathrm{f}}}, \tag{7.3}$$

where τ_in is the round-trip time of the internal laser cavity. Equation (7.3) is only valid if the external feedback reflectance is much smaller than the facet reflectivity, i.e. weak optical feedback.

By inserting the ansatz $E(t) = E_\mathrm{s} e^{\imath\omega_\mathrm{s} t}$ into equation (7.1), relations for the stationary oscillation frequency ω_s and the threshold gain g_s of the laser modes can be found:

$$\omega_\mathrm{s} - \omega_\mathrm{l} = -\frac{C}{\tau_\mathrm{ext}} \sin\left[\arctan(\alpha_\mathrm{H}) + \omega_\mathrm{s}\tau_\mathrm{ext}\right] \tag{7.4}$$

$$g_\mathrm{s} - g_\mathrm{l} = -\frac{2n_\mathrm{g}C}{c_0\tau_\mathrm{ext}\sqrt{1+\alpha_\mathrm{H}^2}} \cos\left[\omega_\mathrm{s}\tau_\mathrm{ext}\right]. \tag{7.5}$$

Here, the feedback parameter C is introduced [137], which is a dimensionless measure of the feedback strength and is defined by

$$C = \tau_\mathrm{ext}\eta_\mathrm{ext}\sqrt{1+\alpha_\mathrm{H}^2}. \tag{7.6}$$

Equation (7.4) may have one or multiple solutions for ω_s, associated with modes called external cavity modes (ECM). In case of multiple solutions, not all ECMs are stable. Based on the derivative of equation (7.4)

$$\partial_{\omega_\mathrm{s}}\omega_\mathrm{l} = 1 + C\cos\left[\arctan(\alpha_\mathrm{H}) + \omega_\mathrm{s}\tau_\mathrm{ext}\right], \tag{7.7}$$

a criterion for the stability is given [139]. For stable ECMs $\partial_{\omega_\mathrm{s}}\omega_\mathrm{l} > 0$ holds.

To better compare the results of the theoretical model and the experimental data, the frequency and gain of the ECMs are expressed in terms of the wavelength shift and relative output power. The wavelength shift with respect to the solitary laser wavelength λ_l is obtained from the frequencies by the approximate linear dependence

$$\Delta\lambda = -\frac{\lambda_\mathrm{l}^2}{2\pi c_0}\left(\omega_\mathrm{s} - \omega_\mathrm{l}\right). \tag{7.8}$$

The relative optical output power P_rel is defined as the ratio

$$P_\mathrm{rel} = \frac{P_\mathrm{out,\,s}}{P_\mathrm{out,\,l}} \tag{7.9}$$

of the output power $P_\mathrm{out,\,s}$ for the stationary state to the power $P_\mathrm{out,\,l}$ for the solitary laser. Since the output power is proportional to the photon density, the output power can be calculated by solving equation (7.2) for $|E|^2$ in the stationary case with $\partial_t N = 0$. The relative optical output power is then given by

$$P_\mathrm{rel} = \frac{g_\mathrm{l}}{g_\mathrm{s}}\left(\frac{\frac{j}{e_0 d_\mathrm{AZ}} - \frac{N_\mathrm{l}}{\tau_N} - \frac{N_\mathrm{s}-N_\mathrm{l}}{\tau_N}}{\frac{j}{e_0 d_\mathrm{AZ}} - \frac{N_\mathrm{l}}{\tau_N}}\right), \tag{7.10}$$

where N_s and N_l are the carrier densities for the stationary and solitary laser state. The latter terms in the main numerator of equation (7.10) are obtained by expanding N_s/τ_N in terms of N_l/τ_N. To find expressions for the carrier density, a linear model for the optical gain g is used (cf. equation (4.13) in section 4.1.1):

$$g = g_\mathrm{diff}(N - N_\mathrm{tr}). \tag{7.11}$$

Here, g_{diff} is the differential gain and N_{tr} is the transparency carrier density. The relation for the relative optical output power then reads

$$P_{\text{rel}} = \frac{g_{\text{l}}}{g_{\text{s}}} \left(\frac{j - (N_{\text{tr}} g_{\text{diff}} + g_{\text{l}}) \frac{e_0 d_{\text{AZ}}}{\tau_N g_{\text{diff}}} - (g_{\text{s}} - g_{\text{l}}) \frac{e_0 d_{\text{AZ}}}{\tau_N g_{\text{diff}}}}{j - (N_{\text{tr}} g_{\text{diff}} + g_{\text{l}}) \frac{e_0 d_{\text{AZ}}}{\tau_N g_{\text{diff}}}} \right). \tag{7.12}$$

If the laser is operated at current densities j well above the threshold, equation (7.12) can be simplified. Noting that the second terms of the numerator and denominator of equation (7.12) equal the threshold current density $j_{\text{th, l}} = e_0 d_{\text{AZ}} N_{\text{l}} / \tau_N$ [62], their contribution is small compared to j. Furthermore, the last term of the numerator is smaller than the threshold current density, even if the gain difference $(g_{\text{s}} - g_{\text{l}})$ is estimated with g_{l} upwards. Therefore, the numerator and denominator are dominated by j and the relative optical output power can be approximated with

$$P_{\text{rel}} = \frac{g_{\text{l}}}{g_{\text{s}}}. \tag{7.13}$$

The approximation (7.13) is not only particularly simple but also allows the relative optical output power to be calculated independently of most material parameters.

Regimes of External Optical Feedback

The influences of external optical feedback on the emission behavior of a diode laser with increasing feedback strength can be roughly classified into five regimes [123], [124].

At the lowest feedback levels, in the feedback regime I with $C < 1$, there is only one solution of equation (7.4) for a given phase $\omega_{\text{s}} \tau_{\text{ext}}$ of the external feedback. With varying external phase, both the frequency ω_{s} and the threshold gain g_{s} of the ECM change periodically relative to the solitary laser. With $C < 1$, the ECM is always stable according to the stability condition based on equation (7.7). In addition, the spectral width of the emission either increases or decreases as a function of the feedback phase.

Feedback regime II is reached when C exceeds unity. Equation (7.4) now has several solutions for the frequency ω_{s}, but not all corresponding ECMs are stable with respect to the stability condition and $C > 1$. Due to the dependence of C, the regime boundary is a function of the external reflector distance. Although certain ECMs in regime II may be stable in terms of the criterion, mode hops between the ECMs may occur. Finally, the laser operates at an ECM with the lowest spectral width of the emission.

If the feedback strength is increased, no more mode hops between ECMs occur, which is associated with regime III.

At even higher feedback levels, additional instabilities occur. The instabilities are caused by the dynamic coupling between optical intensity and carrier density [140], which is characterized by the relaxation oscillation frequency [141]

$$\omega_{\text{r}} = \sqrt{\frac{I - I_{\text{th}}}{(I_{\text{th}} - I_{\text{tr}}) \tau_{\text{p}} \tau_N}}. \tag{7.14}$$

τ_p is the photon lifetime in the internal laser cavity, whereas I, I_{th} and I_{tr} are the injection current at the working point, the threshold current and the transparency current, respectively. While the parameter space for stable ECMs is reduced, the emission behavior becomes more complex. A variety of emission states, like e.g. pulsations, can occur [142], which are characterized by additional frequencies in the emission spectrum. Finally, a totally chaotic emission behavior with an extremely increased spectral width of the emission can be reached. This regime is called regime IV, coherence collapse [127]. The transition from the occurrence of the first instabilities to coherence collapse can be manifested by a complex dynamic behavior with alternating stable and unstable states without a clear boundary [128], [129], [143]. However, a critical feedback strength for the occurrence of coherence collapse can be defined based on the relaxation oscillation frequency. As an upper limit, the critical feedback strength can be estimated with the external coupling rate equal to the relaxation oscillation frequency $\eta_{ext} = \omega_r$ [144]. Depending on the laser properties, the estimated critical feedback reflectance may be larger than the front facet reflectivity of the laser, which is beyond the validity of the Lang-Kobayashi equations. For long distances to the external reflector yielding $\omega_r \tau_{ext} > 1$, typically several dozens of centimeters, regime IV is expected to be reached regardless of the distance and the phase of the external feedback [145], [146]. In contrast, if the distance to the reflector is short, no coherence collapse occurs, especially if the laser is still in regime I with $C < 1$.

The regime V is reached when the laser emission becomes stable again. Here, the configuration is similar to a typical external-cavity diode laser (ECDL).

Exemplary Emission Characteristics with External Optical Feedback

The dependence of the frequency and threshold gain of the ECMs and the associated wavelength shift and relative output power are exemplary depicted in figure 7.2. For the example calculation, a laser with a solitary emission wavelength and a threshold gain of $\lambda_l = 1120\,\text{nm}$ and $g_l = 1000\,\text{m}^{-1}$ and a linewidth enhancement factor of $\alpha_H = 1$ is assumed. The external reflector is located at a distance of $l_{ext} = 0.5\,\text{m}$. The frequency and threshold gain of the ECMs lie on ellipses whose size increases linearly with the feedback parameter. Due to the nonlinear dependence of the relative output power on the threshold gain of the ECM, the relative output power and wavelength shift curves become asymmetric with increasing feedback parameter. Here, the increase in output power due to a lower threshold gain is greater than the decrease in output power due to an increased threshold gain. A variation of the external phase from 0 to 2π is equivalent to a clockwise cycle starting at the position on the ellipse with minimum gain and maximum relative output power, respectively. It is worth noting that the shape and size of the ellipses do not depend on the distance to the external reflector. Consequently, as long as the external feedback reflectance remains constant, the maximum possible shift of the emission wavelength or output power is similar for all external feedback configurations. At feedback levels with $C > 1$, up to half of the ellipses correspond to unstable ECMs. In this case, an ECM with increased threshold gain and reduced optical output power is always associated with a higher frequency and negative wavelength shift.

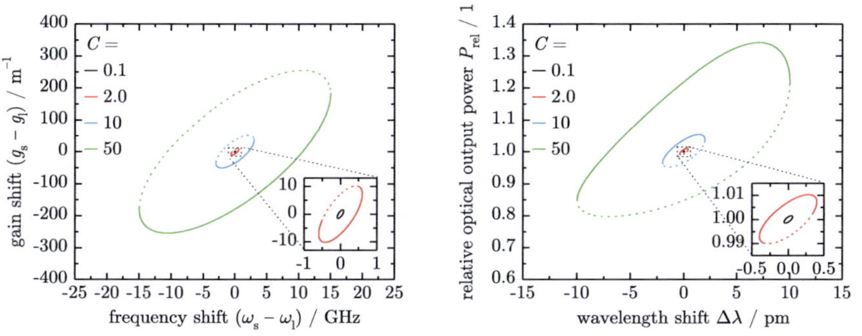

(a) (b)

Figure 7.2.: Exemplary relationship between (a) threshold gain shift and frequency shift and (b) relative optical output power and wavelength shift of possible external cavity modes (ECM) for a laser with different external feedback parameters C. Dotted curves indicate instability of the corresponding ECMs for $C > 1$.

Lasers with Distributed Bragg-Reflector

Originally, the Lang-Kobayashi equations have been derived for single spectral mode emitting Fabry-Pérot lasers. Unlike a plane reflecting facet, a DBR is spatially extended and has strong dispersive properties that must be considered. By appropriate renormalization of the parameters entering the Lang-Kobayashi equations, the properties of the DBR can be taken into account while preserving the original scheme of the approach [141], [147].

In typical application scenarios, DBR-RWLs emit at wavelengths very close to the maximum reflectance of the DBR, i.e. with little detuning. In this case, the effects of detuning can be neglected and the consideration of the DBR in the Lang-Kobayashi equations can be greatly simplified. The description of the DBR is then reduced to the maximum reflectance R_{DBR} and the effective round-trip time in the DBR section τ_{DBR}. The round-trip time corresponds to an effective penetration depth into the grating. Consequently, only the equations for the threshold gain g_{l} of the solitary laser mode and the internal round-trip time τ_{in} need to be adjusted accordingly. The threshold gain is defined by the threshold condition

$$g_{\mathrm{l}} = \alpha_0 - \frac{1}{2l_{\mathrm{g}}} \ln(R_{\mathrm{f}} R_{\mathrm{DBR}}) \tag{7.15}$$

where α_0 is the background loss. Since the actual laser cavity extends into the DBR section, the equation for the internal round-trip is

$$\tau_{\mathrm{in}} = \frac{2n_{\mathrm{g}} l_{\mathrm{g}}}{c_0} + \tau_{\mathrm{DBR}}. \tag{7.16}$$

Finally, the photon lifetime is given by

$$\tau_{\mathrm{p}} = \frac{\tau_{\mathrm{in}}}{2 l_{\mathrm{g}} g_{\mathrm{l}}}. \tag{7.17}$$

Thus, the DBR-RWL can be described similar to a Fabry-Pérot laser with modified back facet properties.

7.2. Investigated Laser Samples

The effects of external optical feedback on the emission characteristics are investigated using DBR-RWLs with different geometries. All DBR-RWL samples are based on the same vertical structure as the SOAs in section 4.2, which is optimized for emission wavelengths around 1120 nm.

In figure 7.3 the schematic of a DBR-RWL is shown. The DBR-RWL consists of a gain

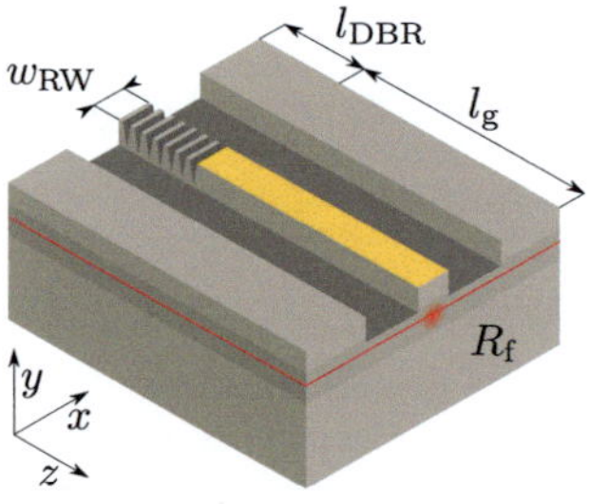

Figure 7.3.: Schematic of a distributed Bragg-reflector ridge waveguide laser. l_{g}: gain section length, l_{DBR}: distributed Bragg-reflector section length, w_{RW}: ridge waveguide width, R_{f}: front facet reflectivity.

section with a length of l_{g} and a distributed Bragg-reflector (DBR) section with a length of $l_{\mathrm{DBR}} = 1\,\mathrm{mm}$ at the back. The DBR is a 6th order grating with a period of 1010 nm and a maximum reflectance of 0.35. Two etched trenches form the ridge waveguide (RW) with a width of $w_{\mathrm{RW}} = 4\,\mu\mathrm{m}$, which extends over the entire DBR-RWL. While the back facet has an antireflection coating to suppress Fabry-Pérot modes, the front facet is coated to realize a specific reflectivity R_{f}.

Both the gain section length l_{g} and the front facet reflectivity R_{f} vary among the investigated DBR-RWLs. The investigated DBR-RWLs have a front facet reflectivity of 0.3 (30 %) and gain section lengths of 1 mm, 2 mm and 3 mm, and additionally front facet reflectivities of 0.01 (1 %) and 0.1 (10 %) for a gain section length of 3 mm, giving a total of five different DBR-RWLs.

The DBR-RWLs are soldered with the epitaxial side up on aluminum nitride heat spreaders. Depending on the gain section length, the DBR-RWLs are designed for injection currents of 0.4 A for the shortest DBR-RWL and more than 1 A for the longest DBR-RWL.

The electro-optical emission characteristics at the default heat sink temperature of 25 °C are shown in figure 7.4a. The emitted optical output power ranges from nearly

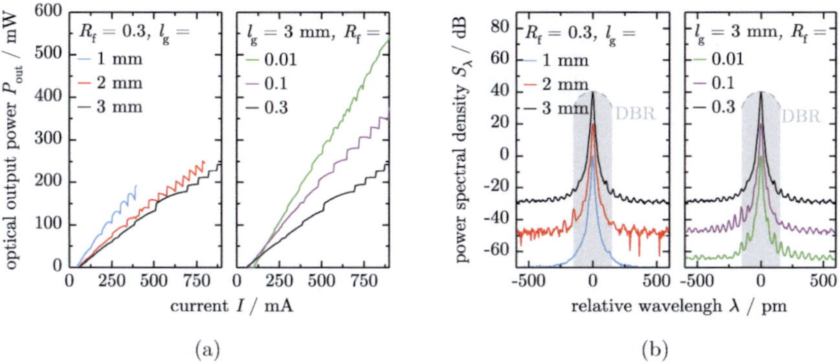

Figure 7.4.: (a) Power-current characteristics of the distributed Bragg-reflector ridge waveguide lasers (DBR-RWL) with different gain section lengths l_g (left) and front facet reflectivities R_f (right). (b) Emission spectra of the DBR-RWLs with different gain section lengths l_g (left) and front facet reflectivities R_f (right) at their default working point with 50 mW optical output power.

200 mW to more than 500 mW at the respective maximum current. Several kinks in the power-current characteristic can be observed, indicating spectral mode hops.

The emission spectra of the DBR-RWLs at the default working point are shown in figure 7.4b. Due to the different internal refractive index distributions resulting from the different geometries and injection currents, the actual emission wavelengths of the five DBR-RWLs differ slightly. Therefore, the emission spectra are plotted as a function of the relative wavelength with respect to the wavelength of maximum intensity. Each DBR-RWL emits radiation in a single spectral mode and side lobes are suppressed by more than 45 dB. In addition to the emission spectra, the reflectance of the DBR grating of the DBR-RWLs, taken from [141], is also shown as a reference. The gray shaded area indicates the region where the reflectance of the grating is greater than 0.1 of the maximum DBR reflectance. Within the typical spectral mode spacing of the individual DBR-RWLs visible in the side-lobes of the emission spectrum, the reflectance of the DBR does not vary much.

7.3. Experimental Setup and Methods

The experimental setup is designed to generate defined levels of external optical feedback to the distributed Bragg-reflector ridge waveguide laser (DBR-RWL). A schematic of the setup is given in figure 7.5. The general configuration is similar to the setup presented in section 6.3. The DBR-RWL is mounted on a heat sink that is actively temperature

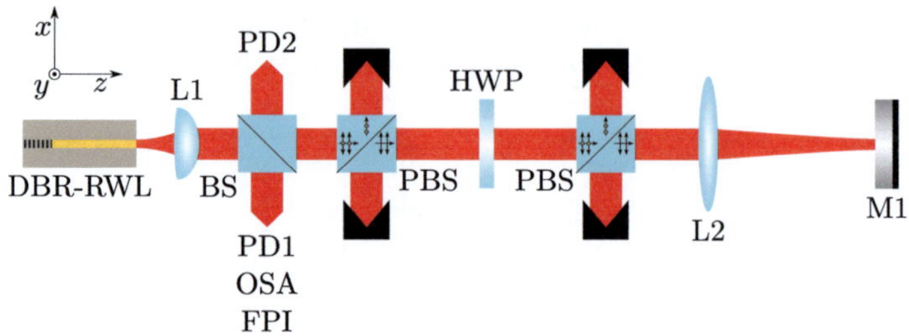

Figure 7.5.: Schematic of the experimental setup for the characterization of distributed Bragg-reflector ridge waveguide lasers (DBR-RWL) with external optical feedback. L1: aspherical lens, BS: beamsplitter, PBS: polarizing beamsplitter, HWP: half wave plate, L2: spherical lens, M1: plane mirror, PD1 / PD2: power detectors, OSA: optical spectrum analyzer, FPI: Fabry-Pérot interferometer.

controlled to a default temperature of 25 °C. The emission from the DBR-RWL is collimated by an aspherical lens L1 with a focal length of 4.5 mm. The output beam passes through a beamsplitter (BS) while a fraction of $R_{\mathrm{BS}} = 0.5$ is directed to various measuring devices. The output power of the DBR-RWL is measured by the power detector PD1 (*XLP12-3S-H2*) and the spectral emission properties are characterized with an optical spectrum analyzer (OSA) and a Fabry-Pérot interferometer (FPI). The polarizing beamsplitters (PBS) and the half wave plate (HWP) are used to variably attenuate the beam. With the combination of the spherical lens L2 with a focal length of 250 mm and the plane silver mirror M1, the beam back is reflected into itself. Finally, M1 is located at a distance of $l_{\mathrm{ext}} = 0.6$ m to the front facet of the DBR-RWL. Similar to the considerations in section 6.3, this configuration results in a high coupling efficiency back into the DBR-RWL. The beamsplitter BS reflects a part of the feedback to the power detector PD2 (*PH20-Ge*), which allows to determine the feedback strength, i.e. the external feedback reflectance. The external feedback reflectance R_{ext} is defined as the ratio of the external feedback power P_{ext} to the output power P_{out} of the DBR-RWL. With the measured powers P_{PD1} and P_{PD2} from power detectors PD1 and PD2, as well as the reflectivity R_{BS} of the beamsplitter BS, the relations for the output power and external feedback power are $P_{\mathrm{out}} = P_{\mathrm{PD1}}/R_{\mathrm{BS}}$ and $P_{\mathrm{ext}} = (1 - R_{\mathrm{BS}})P_{\mathrm{PD2}}/R_{\mathrm{BS}}$, respectively. Therefore, the external feedback reflectance is given by

$$R_{\mathrm{ext}} = (1 - R_{\mathrm{BS}})\frac{P_{\mathrm{PD2}}}{P_{\mathrm{PD1}}}. \tag{7.18}$$

With this setup, external feedback reflectances from 1×10^{-5} to 1.5×10^{-1} can be realized.

A working point around 50 mW optical output power is set as the default for all DBR-RWLs. Since the DBR-RWLs with 3 mm gain section length and facet reflectivities of

0.01 and 0.1 are also used in high-power applications, a second working point is defined, where optical output powers of 510 mW and 350 mW are achieved. To avoid the effects of spectral mode hopping, the actual injection current is increased to a working point far away from a mode hop. This also ensures that the emission wavelength is close to the wavelength of maximum reflectance of the DBR. The default currents are given in table 7.1.

Table 7.1.: Working points of the distributed Bragg-reflector ridge waveguide lasers with different gain section lengths l_g and front facet reflectivities R_f.

gain section length l_g / mm	front facet reflectivity R_f / 1	optical output power $P_{out,1}$ / mW	current I / mA
1	0.3	50.0	131
2	0.3	50.0	179
3	0.3	48.5	207
3	0.01	50.0	161
3	0.1	50.0	165
3	0.01	510	850
3	0.1	350	870

The experimental setup is first aligned with each DBR-RWL to optimize the coupling of the feedback into the DBR-RWL. The feedback is set to the maximum strength and the DBR-RWL is operated at a current just above the laser threshold. The optical elements, and in particular the position of mirror M1, are then aligned to maximize the output power of each DBR-RWL. This corresponds to a configuration where the external cavity mode (ECM) has minimum threshold gain and the increase in optical power is sensitive to the coupling efficiency. After the alignment procedure, the DBR-RWL is set to the default working point and the actual measurements are taken. At each individual feedback strength, the feedback phase is adjusted to the minimum possible optical output power by making tiny adjustments to the mirror position. The corresponding relative optical output power P_{rel} is defined with respect to the output power of the solitary laser:

$$P_{rel} = \frac{P_{PD1,\,s}}{P_{PD1,\,l}}. \tag{7.19}$$

Here, $P_{PD1,\,s}$ and $P_{PD1,\,l}$ are the measured power levels from power detector PD1 for the stationary state with external feedback and the solitary laser without feedback, respectively. In the absence of instability, the mode with the lowest optical output power corresponds to an external feedback phase of π. Due to instability of certain ECMs, the external phase of the mode with the minimum output power may be larger.

The parameters used for the calculations can be found in table B.7.

7.4. Results

First, the influence of the external optical feedback on the relative optical output power is examined while compared to calculations using equation (7.13). In figure 7.6 the relative optical output power of the DBR-RWLs with a front facet reflectivity of 0.3 and different gain section lengths as a function of the external feedback reflectance is given. When the relative optical output power is plotted as a function of the external feedback

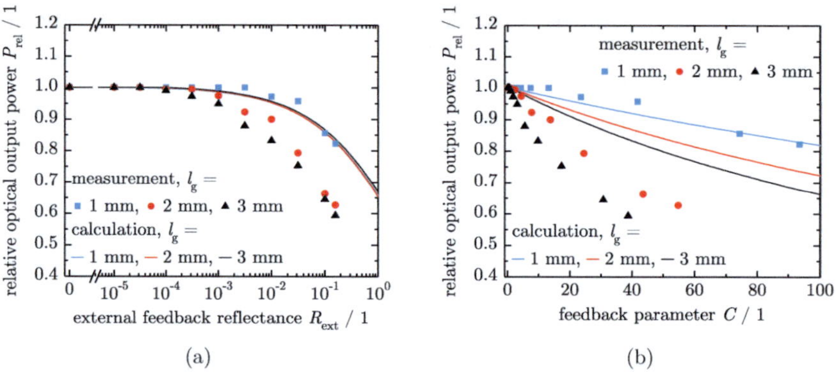

Figure 7.6.: Relative optical output power of distributed Bragg-reflector ridge waveguide lasers with a front facet reflectivity of $R_f = 0.3$ and different gain section lengths l_g at the default working point (a) as a function of the external feedback reflectance and (b) as a function of the feedback parameter.

reflectance, almost the same behavior is observed for the DBR-RWLs with longer gain section lengths. At the highest feedback reflectance, the optical power decreases by 39 % and 41 % for the DBR-RWLs with gain section lengths of 2 mm and 3 mm, respectively. In contrast, the output power of the DBR-RWL with a gain section length of 1 mm decreases by 20 %. The calculations for the stable external cavity mode (ECM) with the lowest relative output power show an almost similar dependence for all gain section lengths. While the calculated results agree with the experimental results for the DBR-RWL with 1 mm gain section length, the power reduction is underestimated for the longer DBR-RWLs. Even for low external feedback reflectances of the order of 1×10^{-3}, the measured power reduction is larger than expected from the calculations.

When the relative optical output power is plotted against the feedback parameter, as in figure 7.6b, the influence of the gain section length on the individual dependence becomes more apparent. Both measured and calculated data show the same qualitative dependence, even though they do not directly match. Based on the feedback parameter, the DBR-RWL with the longest gain section length is most affected by power loss.

In figure 7.7 the relative output power for DBR-RWLs with a gain section length of 3 mm and different front facet reflectivities is shown. Compared to the DBR-RWL

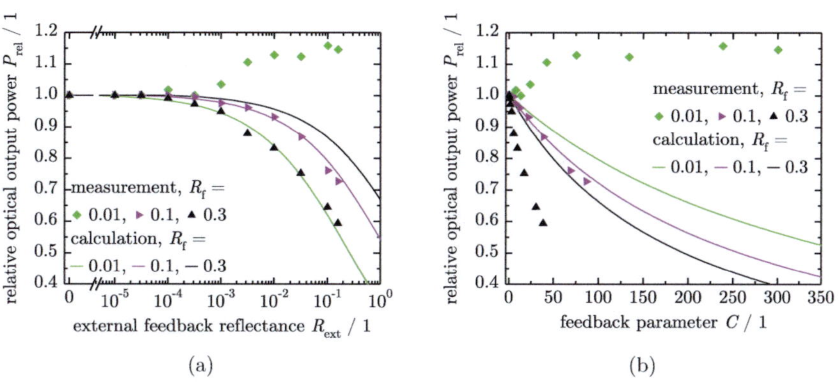

Figure 7.7.: Relative optical output power of distributed Bragg-reflector ridge waveguide lasers with a gain section length of $l_g = 3\,\text{mm}$ and different front facet reflectivities R_f at the default working point (a) as a function of the external feedback reflectance and (b) as a function of the feedback parameter.

with a front facet reflectivity of 0.3, the power decrease of the DBR-RWL with a front facet reflectivity of 0.1 is smaller. At the highest external feedback reflectance, the output power decreases by 27 %. A completely different behavior is observed for the DBR-RWL with a front facet reflectivity of 0.01. While the output power is not significantly reduced, the relative output power actually increases at high external feedback reflectances. Therefore, it is likely that the ECMs with reduced output power are not stable for the DBR-RWL with a front facet reflectivity of 0.01. In the calculated relative optical output powers, a clear influence of the front facet reflectivity can be observed both in the plot as a function of the external feedback reflectance and as a function of the feedback parameter. However, good agreement with the measurement is found only for the laser with a front facet reflectivity of 0.1.

Next, the effects on the emission wavelength of the DBR-RWLs at varying external feedback reflectances are investigated. The experimental results are compared to calculations based on equation (7.4). In figure 7.8a the wavelength shift as a function of the external feedback reflectance for the lasers with a front facet reflectivity of 0.3 is given. The observed wavelength shift at high external feedback reflectances is positive for all three DBR-RWLs.

The two DBR-RWLs with gain section lengths of 1 mm and 2 mm show comparable wavelength shifts, with a slightly larger maximum shift of 40 pm observed for the DBR-RWL with a gain section length of 1 mm. Furthermore, the wavelength shift of the laser with a gain section length of 3 mm is characterized by a large shift of 60 pm at an external feedback reflectance of 3×10^{-3}. With respect to the internal cavity spectral mode spacing of 51 pm for the laser with 3 mm gain section length, the wavelength shift can be identified as a spectral mode hop. The experimental data are in contradiction

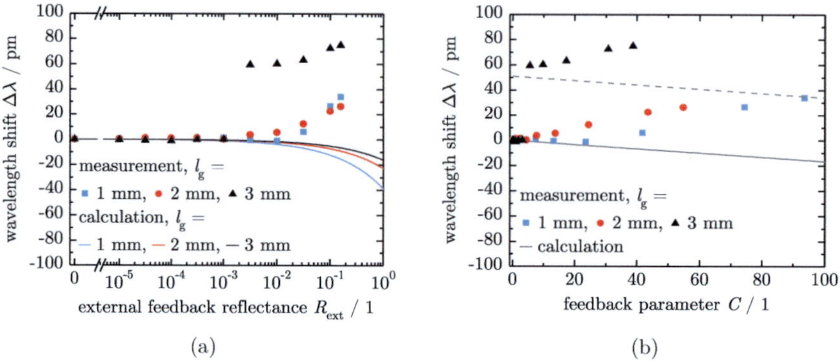

Figure 7.8.: Emission wavelength shift of distributed Bragg-reflector ridge waveguide lasers with a front facet reflectivity of $R_f = 0.3$ and different gain section lengths l_g at the default working point (a) as a function of the external feedback reflectance and (b) as a function of the feedback parameter.

with the calculation, where only a negative wavelength shift is obtained for the stable ECM with the lowest output power. In addition, the calculated absolute wavelength shift is at least a factor of two smaller than the measured one.

In the plot of the wavelength shift as a function of the feedback parameter (figure 7.8b), all DBR-RWLs should show the same behavior. Therefore, the calculated wavelength shift is plotted as a single curve. Additionally, the calculated wavelength shift for the next red-shifted mode of the laser with a gain section length of 3 mm is shown. Not only the sign of the wavelength shift but also the individual dependence of the absolute wavelength shift on the feedback parameter is inconsistent with the calculation.

The comparison of DBR-RWLs with different front facet reflectivities and a gain section length of 3 mm is presented in figure 7.9. Similar to the previous results, the wavelength shift at high external feedback reflectances is almost always positive. An exception is observed for the DBR-RWL with a front facet reflectivity of 0.01. At an external feedback reflectance of 1×10^{-1}, the wavelength decreases by more than 40 pm. With respect to the spectral mode spacing of the DBR-RWLs with a gain section length of 3 mm, the shift can also be related to a mode hop. For the DBR-RWL with a front facet reflectivity of 0.1, a slight negative wavelength shift is observed before increasing to a maximum red-shift around 20 pm.

In figure 7.9b the wavelength shift as a function of the feedback parameter is given. Again, a different behavior is obtained for all three DBR-RWLs, while the calculated wavelength shift shows the same linear dependence as before. Only for the laser with a front facet reflectivity of 0.1 there is a partial agreement for low feedback parameters.

The measured power decrease and the wavelength shift, especially for DBR-RWLs with gain section lengths above 1 mm, do not agree with the calculated results. Even for

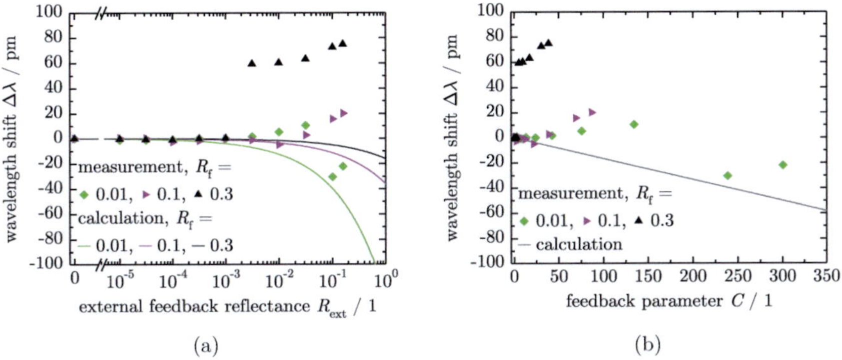

Figure 7.9.: Emission wavelength shift of distributed Bragg-reflector ridge waveguide lasers with a gain section length of $l_g = 3\,\text{mm}$ and different front facet reflectivities R_f at the default working point (a) as a function of the external feedback reflectance and (b) as a function of the feedback parameter.

relatively low external feedback reflectances, large deviations are found. Considering the parameter space for stable ECMs, there are no modes that lead to the measured results.

However, the behavior can be explained in terms of the power decrease. As less power is dissipated from the DBR-RWL via the emission of laser radiation, more waste heat is generated. Although the DBR-RWLs are mounted on an actively controlled heat sink, a finite thermal resistance results in an increase of the local temperature. On the one hand, the thermo-optical effect in the internal laser cavity leads to a red-shift of the current lasing ECM. On the other hand, the reflection characteristics of the DBR are also affected by the thermo-optical effect and the reflectance maximum also experiences a red-shift with increasing temperature. However, these thermal effects are not included in the theoretical model.

The spectral emission behavior of the DBR-RWLs is further characterized by means of the spectral width. Dynamic emission states, such as coherence collapse, can thus be distinguished from emission states with stable ECMs. Additionally, the results are evaluated with respect to the critical feedback strength based on equation (7.14). In figure 7.10 the spectral width for DBR-RWLs with different gain section lengths and a facet reflectivity of 0.3 is shown. The critical feedback strength is symbolized by vertical lines. The measured spectral width is around 3 MHz with only a few exceptions. With respect to the resolution limit of the Fabry-Pérot interferometer, it can be assumed that this limits the measurement and that the actual spectral width could be even smaller. For the DBR-RWL with a gain section length of 2 mm, a slightly increased spectral width at an external feedback reflectance of 1×10^{-5} is determined. In contrast, the spectral width for the DBR-RWL with a gain section length of 1 mm is about 7 MHz in solitary operation without external optical feedback. The spectral width decreases

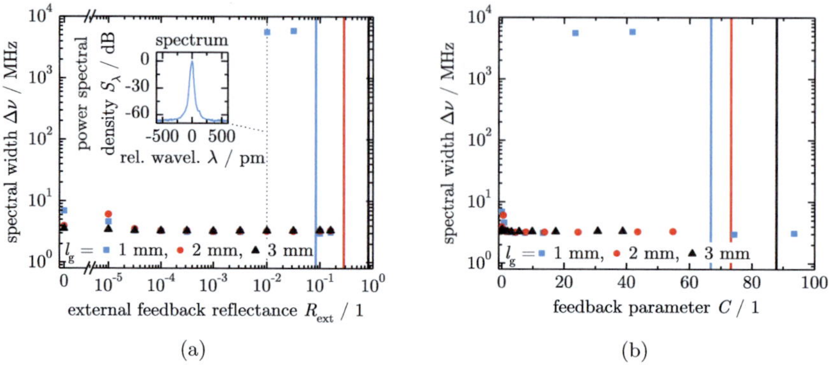

Figure 7.10.: Spectral width of the emission of distributed Bragg-reflector ridge waveguide lasers (DBR-RWL) with a front facet reflectivity of $R_f = 0.3$ and different gain section lengths l_g at the default working point (a) as a function of the external feedback reflectance and (b) as a function of the feedback parameter. Critical feedback levels are marked by vertical lines for each DBR-RWL. The inset shows the emission spectrum of the DBR-RWL with a gain section length of 1 mm at an external feedback reflectance of $R_{ext} = 1 \times 10^{-2}$, corresponding to a feedback parameter of $C = 23.5$.

with increasing feedback reflectances. At external feedback reflectances of 1×10^{-2} and 3×10^{-2}, however, the spectral width increases by several orders of magnitude to nearly 6000 MHz. The corresponding emission spectrum at an external feedback reflectance of 1×10^{-2} and a feedback parameter of 23.5 is shown in the inset of figure 7.10a. As in solitary laser operation, the emission spectrum is characterized by a central lobe, but the width is measurably increased.

The effects at low external feedback reflectances are consistent with the typical increase or decrease in spectral width in the feedback regimes I and II, especially when the feedback parameter is close to unity. For the DBR-RWL with a gain section length of 1 mm, the transition to regime IV is visible by the coherence collapse with a drastically increased spectral width. At feedback reflectances of 1×10^{-1} and higher, the again small spectral width indicates the feedback regime V.

Compared to the calculated critical feedback strength, the coherence collapse is already observed for external feedback reflectances that are one order of magnitude smaller. In contrast, the calculation for DBR-RWLs with longer gain section lengths yields higher critical feedback strengths outside the investigated range of feedback strengths. Here, the absence of coherence collapse in the measurements is in agreement with the calculations.

The comparison of the spectral width for DBR-RWLs with a gain section length of 3 mm and different front facet reflectivities is shown in figure 7.11. Again, for most of the measurements, a spectral width near 3 MHz is obtained for all DBR-RWLs. For the

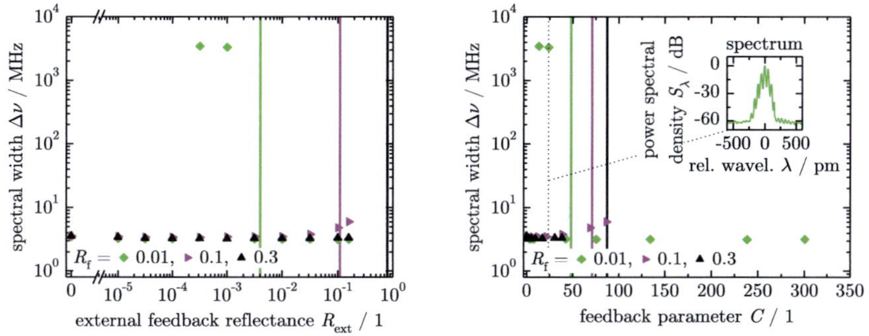

(a) (b)

Figure 7.11.: Spectral width of the emission of distributed Bragg-reflector ridge waveguide lasers with a gain section length of $l_g = 3\,\text{mm}$ and different front facet reflectivities R_f at the default working point (a) as a function of the external feedback reflectance and (b) as a function of the feedback parameter. Critical feedback levels are marked by vertical lines for each laser type. The inset shows the emission spectrum of the DBR-RWL with a front facet reflectivity of 0.01 at a feedback parameter of 23.9, corresponding to an external feedback reflectance of $R_{ext} = 1 \times 10^{-3}$.

DBR-RWL with a front facet reflectivity of 0.01, the measured spectral width raises to more than 3000 MHz at external feedback reflectances of 3×10^{-4} and 1×10^{-3}. For both external feedback reflectances a similar emission spectrum is obtained. In the inset of figure 7.11b the emission spectrum of the DBR-RWL at an external feedback reflectance of 1×10^{-3} and a corresponding feedback parameter of 23.9 is given. Compared to solitary laser operation, the emission spectrum has changed fundamentally. On the one hand, the width of the central lobe has increased. On the other hand, several side lobes with an equal spacing around 50 pm are observed. Such emission spectra may indicate pulsations [142], rather than coherence collapse. For a clear differentiation, further measurements, such as the time trace of the optical output power, are necessary. While no coherence collapse occurred for the DBR-RWL with a front facet reflectivity of 0.1, a small increase in spectral width can be observed at the highest external feedback reflectances.

Although coherence collapse is not observed, the calculated critical feedback strengths agree with the measured data and give upper estimates for stable ECM emission states. Especially noticeable is the critical feedback strength for the DBR-RWL with a front facet reflectivity of 0.1. This value is close to the front facet reflectivity and at the limit of the investigated range of external feedback reflectances. Therefore, the experimentally observed increased spectral widths may indicate the onset of coherence collapse.

Finally, the DBR-RWLs with front facet reflectivities of 0.01 and 0.1 are investigated at the high-power working points. In figure 7.12a the corresponding relative optical output power is shown as function of the external feedback reflectance. The relative

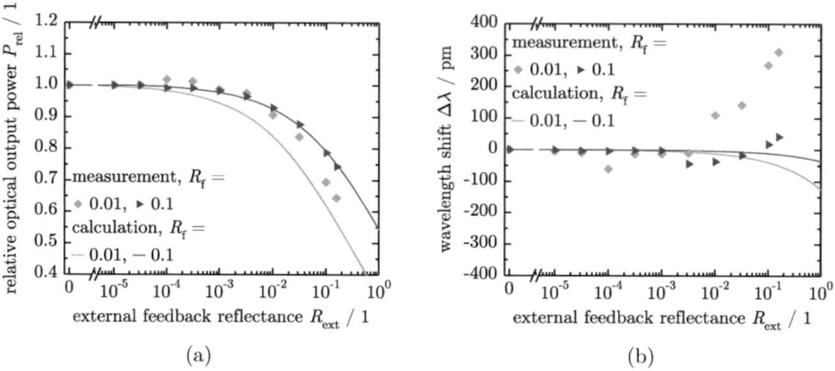

Figure 7.12.: (a) Relative optical output power and (b) emission wavelength shift of distributed Bragg-reflector ridge waveguide lasers with a gain section length of $l_g = 3\,\text{mm}$ and different front facet reflectivities R_f at the high-power working point as a function of the external feedback reflectance.

output power of the DBR-RWL with a front facet reflectivity of 0.01 increases slightly at low external feedback reflectances of the order of 1×10^{-4}. At higher external feedback reflectances, the output power decreases strongly by up to 35 %. For the other DBR-RWL only a continuously decreasing relative output power with a maximum decrease of 25 % is observed. The calculations agree with the measured data, with particularly good agreement obtained for the DBR-RWL with a front facet reflectivity of 0.1. Thus, the ECMs with the lowest optical output power are not limited by further instabilities in the case of the high-power working point.

The emission wavelength shift is shown in figure 7.12b. Again, strongly varying central emission wavelengths and no agreement with the calculation are observed. On the one hand, there is a red-shift at large external feedback reflectances. Especially the laser with a front facet reflectivity of 0.01 is characterized by a significantly large shift of more than 300 pm. On the other hand, several mode hops can be distinguished, which first appear at external feedback reflectances of 1×10^{-4} and 3×10^{-3} for the DBR-RWLs with front facet reflectivities of 0.01 and 0.1, respectively. With respect to the high-power working point, relative changes in optical output power result in higher amounts of waste heat. Thus, the thermally driven wavelength shifts are much larger.

An emission state with increased spectral width is only observed for the DBR-RWL with a front facet reflectivity of 0.01 at an external feedback reflectance of 1×10^{-2}. The emission spectrum is similar to the low-power working point (see inset of figure 7.10b) and pulsations can be assumed. The calculated critical feedback strength is shifted towards higher external feedback reflectances for the high-power working points. Consequently, the calculations yield critical external feedback reflectances of 4.4×10^{-2} and 8.5×10^{-1}

for the DBR-RWLs with front facet reflectivities of 0.01 and 0.1, respectively. Although still overestimated, the calculations are in agreement with the experimental data.

7.5. Summary

In this chapter, the emission behavior of distributed Bragg-reflector ridge waveguide lasers (DBR-RWL) with an emission wavelength of 1120 nm has been characterized under the influence of external optical feedback. Five different DBR-RWLs have been used to study the effects of both gain section length, front facet reflectivity, and the particular working point. For external optical feedback reflectances up to 1.5×10^{-1}, the optical output power and the central emission wavelength can be subject to significant changes. Even in the absence of coherence collapse, the optical output power can decrease by several ten percent and wavelength shifts by dozens of picometers can occur. Calculations based on an adaptation of the Lang-Kobayashi equations to DBR-RWLs show only partial agreement. It turns out that thermal effects inside the DBR-RWLs have a strong influence on their emission behavior under external optical feedback, which is not covered by the theoretical model.

For DBR-RWLs with a constant front facet reflectivity of 0.3 but varying gain section length, opposite trends in relative output power and wavelength shift occur. On the one hand, with increasing gain section length, the possible output power reduction increases from 20 % to 40 % at the highest external feedback reflectance. First significant power changes are already visible at external feedback reflectances of 3×10^{-4}. On the other hand, the maximum wavelength shift decreases with the gain section length. Additionally, spectral mode hops corresponding to the internal cavity can significantly increase the wavelength shift to more than 50 pm. Due to their smallest spectral mode spacing, mode hops are more likely for the lasers with a gain section length of 3 mm. At a default working point with 50 mW output power, coherence collapse is found only for the laser with the shortest gain section length of 1 mm at external feedback reflectances of 1×10^{-2} and higher.

Concerning DBR-RWLs with varying front facet reflectivity but constant gain section length of 3 mm, the influence of the front facet reflectivity is ambivalent. Especially for low facet reflectivities of 0.01, the actual working point has a strong influence. At the default working point with 50 mW optical output power, the maximum possible power loss at high external feedback reflectances increases with increasing front facet reflectivity. Here, the behavior of the DBR-RWL with a front facet reflectivity of 0.01 is affected by the appearance of additional instabilities. With external feedback reflectances exceeding 1×10^{-4}, emission spectra with increased spectral widths and multiple side lobes are observed, possibly indicating pulsations. There is no clear trend between the facet reflectivities in the emission wavelength shift. The wavelength shift is dominated by thermal effects and mode hops and reaches values of 20 pm and more.

The working point has a significant influence on the emission behavior under external optical feedback for DBR-RWLs with a low facet reflectivity of 0.01. When operating at a working point with several hundred milliwatts of optical output power, the highest reduction in output power is found for the lowest facet reflectivity. On the one hand, the parameter space for stable modes is not significantly limited by instabilities and the critical feedback strength is shifted to higher external feedback reflectances. For the DBR-RWL with a front facet reflectivity of 0.01 possible pulsations occur only at a relatively high external feedback reflectance of 1×10^{-2}. On the other hand, the thermally induced wavelength shift and spectral mode hops lead to much larger emission wavelength fluctuations of several hundred picometers.

The calculations based on a steady-state approach of the Lang-Kobayashi equations are in partial agreement with the measurements. In most cases, the calculation can be used to explain the measured power loss well, at least as a rough estimate. In contrast, the calculated wavelength shift is not in agreement with the measurement, since the dominant thermal and mode-hopping effects are not included in the theory. However, the occurrence of coherence collapse and the corresponding dependence on the working point are in good agreement with the calculations.

8. Conclusions of Part II.

In this part, the effects of external optical feedback on the operational behavior of tapered amplifiers (TPA) and distributed Bragg-reflector ridge waveguide lasers (DBR-RWL) have been analyzed experimentally and theoretically. For TPAs, external feedback reflectances up to 10^{-2} slightly reduce the output power, but increase the feedback emission of the TPA itself by orders of magnitude. DBR-RWLs can also experience power changes, with a decrease of nearly half observed for external feedback reflectances of 1.5×10^{-1}. In addition to the chaotic emission behavior in the coherence collapse regime, thermally driven emission wavelength changes and mode hops lead to emission wavelength shifts of several hundred picometers.

The effect of external optical feedback on the operating behavior of TPAs depends primarily on the coupling of the feedback into the ridge waveguide (RW). In typical applications, where the output emission of the TPA is coupled into periodically poled nonlinear crystals (PPNC) or fibers, the feedback beam to the TPA is generated by a reflection into itself. Therefore, this ideally matched external feedback, has the greatest influence on the emission behavior. The shape of the feedback beam may differ due to the working point dependent astigmatism or specific spatial emission characteristics of e.g. bulk PPNCs. Here, the external optical feedback is usually less critical than the matched feedback. The output emission of the TPA is only weakly dependent on the external feedback and changes are only noticeably for external feedback reflectances greater than 1×10^{-3}. The spatial emission characteristics of the output beam remain almost unchanged, with the power loss mainly affecting the central beam lobe. In contrast, the feedback emission of the TPA is very sensitive to external optical feedback. Already for external feedback reflectances larger than 1×10^{-5}, the TPA feedback emission starts to increase. For external feedback reflectances of 1×10^{-2}, the feedback emission can exceed the incident seed power level by almost an order of magnitude. Again, the spatial emission characteristics change little, and the feedback emission of the TPA is still determined by the emission of the RW. Thus, the intense and directed feedback emission of the TPA under external optical feedback has a large overlap with the incident seed light. Due to the high sensitivity of the feedback emission of the TPA, external optical feedback must always be considered when designing a master oscillator power amplifier system.

The effects of external optical feedback on the emission properties of DBR-RWLs vary greatly with the resonator geometry of the laser. On the one hand, lasers with short resonator lengths and low front facet reflectivities exhibit emission states with increased spectral width like coherence collapse, which can occur at external feedback reflectances larger than 1×10^{-4}. On the other hand, the output power of DBR-RWLs with front facet reflectivities higher than 0.1 and resonators longer than 1 mm can decrease by up to 40 %

at external feedback reflectances of 1.5×10^{-1}. Thermal effects in the DBR-RWL, also caused by the output power reduction and increasing with higher power working points, result in emission wavelength shifts of 20 pm and more for all DBR-RWLs. Furthermore, especially lasers with long resonator lengths suffer from spectral mode hops that increase the spectral shift by more than 50 pm. In hybrid integrated diode laser systems, the typical distances to the feedback source are in the range of millimeters to centimeters. If the optical feedback is generated outside the hybrid diode laser system, distances of the order of meters are also possible. Since coherence collapse occurs especially for long distances to the feedback source, it may not necessarily be observed in hybrid integrated diode laser systems. In contrast, the reduction of the optical output power and thermal effects are independent of the distance to the feedback source. Therefore, when evaluating the resistance of DBR-RWLs to external optical feedback, not only the occurrence of coherence collapse must be taken into account. Depending on the actual laser source requirements, a compromise must be made when selecting a DBR-RWL for a specific application with unavoidable optical feedback.

9. Final Conclusions and Outlook

In this work, the feedback aspects of periodically poled nonlinear crystals (PPNC), semiconductor optical amplifiers (SOA), and distributed Bragg-reflector ridge waveguide lasers (DBR-RWL), the typical components of hybrid integrated diode laser systems, have been systemically investigated. According to the typical setup of the individual components, the feedback emission characteristics and the effects of optical feedback on the operating behavior have been investigated both experimentally and theoretically. Thus, conclusions for the combination in hybrid integrated diode laser systems can be drawn from the investigation of the individual components.

PPNCs generate optical feedback with reflectances around 10^{-4} to 10^{-3}, which is directed towards the laser source. In addition, resonances occur at certain wavelengths because the periodic poling acts as a Bragg-grating, resulting in reflectances greater than 10^{-2}. In combination with DBR-RWLs, even the low reflectance away from the resonances can negatively affect the operating performance. To achieve high nonlinear conversion efficiencies, high optical power is required and typically DBR-RWLs with longer resonators and low front facet reflectivities are used. On the one hand, operating at high output power supports stable emission states at the given feedback strengths. On the other hand, the emission wavelength can still change by dozens of picometers due to thermal effects and especially due to spectral mode hops. As a result, phase matching is no longer achieved and stable and reliable operation can not be guaranteed. Since DBR-RWLs with shorter resonator lengths tend to be less susceptible to spectral mode hops, these should be taken into account when selecting components.

The optical feedback emitted by SOAs has special characteristics due to the fact that SOAs are active devices. Thus, spontaneous emission can lead to feedback emission even without the introduction of seed light. In addition, the feedback emission typically increases with increasing output power. With respect to the typical incident seed power, this results in reflectances in the range of 10^{-2} to 10^{-1}. The ridge waveguide (RW) structure in the investigated SOAs dominates the spatial emission characteristics, resulting in a highly directed optical feedback towards the seed laser source. DBR-RWLs with high front facet reflectivities are typically used to provide of the required seed power. Even with the high reflectances of the SOAs, operation without coherence collapse is possible. However, significant power reduction and emission wavelength shifts can occur, limiting the use of the master oscillator power amplifier (MOPA) as a light source for certain applications. To minimize the feedback emission of the SOAs, they should always be operated at sufficiently high seed powers in the saturation region. Increasing the seed power beyond saturation can also reduce feedback effects, especially if no optical isolator is available. On the one hand, the feedback emission is reduced in absolute terms. On the other hand, even if the feedback emission remains unchanged, the reflectance of the

SOA is reduced, so that the relative feedback on the DBR-RWL is lower.

When MOPA systems based on SOAs are combined with additional feedback sources, the main effect is increased feedback emission from the SOA. With the external feedback reflectances of PPNCs, the feedback emission of the SOA can increase by an order of magnitude and even exceed the incident seed power level. Thereby the general spatial emission characteristics of the output and the feedback emission are almost unaffected. This is also true even for differing beam shapes of the feedback beam, since the effects depend mainly on the back coupling of the feedback into the RW of the SOA. With the strong feedback emission of the SOA, this must also be considered when using optical isolators. On the one hand, the maximum isolation has to be selected accordingly. On the other hand, the feedback power is so high that the damage threshold of the isolator is relevant.

By knowing the detailed feedback effects in hybrid integrated diode laser systems, they can be taken into account in the design and the individual components can be selected accordingly. In this way, the systems can be further optimized and additional applications can be accessed. However, there are also starting points for the optimization of the individual components.

The characterization of the reflectance spectrum of PPNCs can be used as a method to determine the quality of PPNCs. The characteristics of the spectrum are particularly sensitive to tolerances of the poling period or waveguide homogeneity. The characterizations can thus be a starting point for further optimization of the PPNC with respect to optical feedback, even if the PPNC quality is already ideal with respect to nonlinear frequency conversion. For example, by reducing the duty cycle error of the periodic poling, the reflectance at wavelengths between the resonances can be reduced.

In contrast to the reduction of the optical feedback, it can also be used constructively in the realization of diode laser systems with nonlinear frequency conversion. In particular, the spectrally very narrow resonances in the reflectance spectrum of PPNCs provide ideal conditions for external wavelength stabilization of diode lasers. Although the quasi phase matching wavelength and the position of the resonances do not correlate, there are certain wavelength ranges where both wavelengths overlap, e.g. around 1178 nm. Further optimization of the crystal properties could extend the possible wavelength ranges. A patent with the international publication number WO2020104519 has been filed as a result of these considerations.

The optical feedback of SOAs with waveguide structure is always emitted according to the emission characteristics of the waveguide, even if there are angular tolerances in the coupling of the seed light. Although the total optical feedback increases, intentional tilted coupling can prevent the feedback of the SOA from returning to the laser. This is especially true when exploiting the limited numerical aperture of microlenses. The optimization is also valid in cases where the SOA itself receives external optical feedback.

Finally, the DBR-RWL can be further optimized to withstand higher amounts of optical feedback. Instead of preventing coherence collapse as in typical approaches, the occurrence of spectral mode hops must also be reduced.

A. Additional Relations

A.1. Reflectance of Weak Bragg-Gratings

The reflection properties of Bragg-gratings formed by periodically arranged partially reflecting interfaces are described in [62]. Based on a generally exact transmission matrix approach, a simplified description is derived for the special case of weakly reflecting Bragg-gratings, which is referred to as the Fourier limit. The interfaces are assumed to have a low reflectivity and only one reflection at each interface is considered. The resulting equation for the amplitude reflectance r_g of the Bragg-grating reads (equation (A7.33) in [62])

$$r_\mathrm{g} = 2r\cos(\delta\Lambda 2)e^{-\imath\delta\Lambda/2}\frac{\tan(m\delta\Lambda)}{\tan(\delta\Lambda)}\frac{1+\imath\tan(\delta\Lambda)}{1+\imath\tan(m\delta\Lambda)}. \tag{A.1}$$

r is the amplitude reflectivity of the interfaces, Λ is the period length of the grating and m is the total number of periods. One period contains two interfaces, one of which can generally introduce a phase jump of π. The relative wave vector δ is defined as

$$\delta = k - 2\pi n/\lambda_\mathrm{r} \tag{A.2}$$

where k is the wave vector, n is the (average) refractive index and λ_r is the resonance wavelength.

In the following, it will be shown that the model given in section 3.1 yields the same results as the description in the Fourier limit. First, the trigonometric functions in equation (A.1) are expressed by their exponential representations:

$$r_\mathrm{g} = 2r\frac{1}{2}\left(e^{\imath\delta\Lambda/2}+e^{-\imath\delta\Lambda/2}\right)e^{-\imath\delta\Lambda/2}\frac{\frac{e^{\imath m\delta\Lambda}-e^{-\imath m\delta\Lambda}}{\imath\left(e^{\imath m\delta\Lambda}+e^{-\imath m\delta\Lambda}\right)}}{\frac{e^{\imath\delta\Lambda}-e^{-\imath\delta\Lambda}}{\imath\left(e^{\imath\delta\Lambda}+e^{-\imath\delta\Lambda}\right)}}\frac{1+\imath\frac{e^{\imath\delta\Lambda}-e^{-\imath\delta\Lambda}}{\imath\left(e^{\imath\delta\Lambda}+e^{-\imath\delta\Lambda}\right)}}{1+\imath\frac{e^{\imath m\delta\Lambda}-e^{-\imath m\delta\Lambda}}{\imath\left(e^{\imath m\delta\Lambda}+e^{-\imath m\delta\Lambda}\right)}}. \tag{A.3}$$

Partial calculation of the multiplications gives

$$r_\mathrm{g} = r\left(1+e^{-\imath\delta\Lambda}\right)\frac{e^{\imath m\delta\Lambda}-e^{-\imath m\delta\Lambda}}{e^{\imath\delta\Lambda}-e^{-\imath\delta\Lambda}}\frac{e^{\imath\delta\Lambda}+e^{-\imath\delta\Lambda}+e^{\imath\delta\Lambda}-e^{-\imath\delta\Lambda}}{e^{\imath m\delta\Lambda}+e^{-\imath m\delta\Lambda}+e^{\imath m\delta\Lambda}-e^{-\imath m\delta\Lambda}}, \tag{A.4}$$

which can be further reduced to

$$r_\mathrm{g} = r\left(1+e^{-\imath\delta\Lambda}\right)\frac{1-e^{-\imath 2m\delta\Lambda}}{1-e^{-\imath 2\delta\Lambda}}. \tag{A.5}$$

By using the identity $1 - e^{-\imath 2\delta\Lambda} = (1 - e^{-\imath\delta\Lambda})(1 + e^{-\imath\delta\Lambda})$ and extracting a factor of $e^{\imath\delta\Lambda}$ the equation simplifies to

$$r_{\mathrm{g}} = re^{\imath\delta\Lambda}\frac{e^{-\imath 2m\delta\Lambda} - 1}{1 - e^{\imath\delta\Lambda}}. \tag{A.6}$$

While this result is still equivalent to equation (A.1), the following transformations are only valid if the two interfaces within each period are characterized by the same reflectivity and do not differ by a phase jump of π. Thus, the phase difference of two reflections is only determined by their geometric path difference. At the resonance wavelength, the phase difference is consequently integer multiples of 2π, leading to constructive interference. Inserting equation (A.2) into (A.6) leads to

$$r_{\mathrm{g}} = re^{\imath k\Lambda}\frac{e^{-\imath 2mk\Lambda} - 1}{1 - e^{\imath k\Lambda}} \tag{A.7}$$

since exponents of the form $2\pi n\Lambda/\lambda_{\mathrm{r}}$ are an integer multiple of 2π and vanish.

To compare the result with the relations derived in section 3.1, the orientation of the grating structures have to be considered in detail. In this work, the first reflecting interface is placed after a distance of $\Lambda_{\mathrm{QPM}}/2$ to the back facet. In contrast, the grating in [62] begins immediately with a reflecting interface. Thus, including the path difference in front of the actual grating in the reflectance leads to an additional factor of $\exp(-\imath k\Lambda)$. With the field amplitude E_{in} of an incident wave, the field amplitude E_{FB} of the corresponding reflected wave is given by

$$E_{\mathrm{FB}} = E_{\mathrm{in}}e^{-\imath k\Lambda}r_{\mathrm{g}}, \tag{A.8}$$

which finally yields

$$E_{\mathrm{FB}} = rE_{\mathrm{in}}\frac{e^{-\imath 2mk\Lambda} - 1}{1 - e^{\imath k\Lambda}}. \tag{A.9}$$

With $r \equiv r_{\mathrm{DW}}$, $\Lambda \equiv \Lambda_{\mathrm{QPM}}$, and noting that the number of DW is twice the number of periods $2m \equiv M$, equation (A.9) is equivalent to equation (3.9) in section 3.1 evaluated at $z = 0$.

A.2. Statistical Description of Periodic Poling Errors

The influence of poling errors of periodically poled nonlinear crystals (PPNC) is analyzed in detail in [30]. Analytical expressions for the ensemble average of the nonlinear conversion efficiency are derived, distinguishing separately the duty cycle error and the period error types separately.

With slight modifications, the reported mathematical scheme can still be used to describe the influences of perturbed poling with a combination of both error types. First, the ensemble average of the reflectance of PPNCs is derived in detail. Second, the ensemble average of the nonlinear conversion efficiency is given.

A.2.1. Reflectance of the Periodic Poling

The ensemble average of the reflectance $\langle R_{\text{PP, plane}}\rangle$ of the periodic poling for plane waves is defined as (see equation (3.19)):

$$\langle R_{\text{PP, plane}}\rangle = |r_{\text{DW}}|^2 \left\langle \left| \sum_{p=1}^{M} e^{-\imath k \Delta z_p} \right|^2 \right\rangle. \tag{A.10}$$

r_{DW} is the amplitude reflectivity of a domain wall (DW), k is the wave vector and M is the total number of DWs. The geometric path difference of the reflection at the pth DW given by (see equation (3.20))

$$\Delta z_p = p\Lambda_{\text{QPM}} + 2\delta_{\text{DE},p} + 2\sum_{q=1}^{p} \delta_{\text{PE},q}. \tag{A.11}$$

The poling period Λ_{QPM} contains two DWs with equal reflection properties. The duty cycle error and the period error are manifested by the deviation of the DW position by δ_{DE} and δ_{PE}, respectively.

First, the calculation of the absolute square in equation (A.10) is expressed as a double sum

$$\left| \sum_{p=1}^{M} e^{-\imath k \Delta z_p} \right|^2 = \sum_{p=1}^{M} \sum_{p'=1}^{M} e^{-\imath k (\Delta z_p - \Delta z_{p'})}. \tag{A.12}$$

For $p = p'$ the exponent vanishes and the sum can be partially calculated while the remaining sum is reorganized:

$$\sum_{p=1}^{M} \sum_{p'=1}^{M} e^{-\imath k (\Delta z_p - \Delta z_{p'})} = M + \sum_{p=1}^{M-1} \sum_{p'=p+1}^{M} e^{-\imath k (\Delta z_p - \Delta z_{p'})} + e^{\imath k (\Delta z_p - \Delta z_{p'})}. \tag{A.13}$$

Inserting the result into equation (A.10) and replacing p' by $j + p$ gives

$$\langle R_{\text{PP, plane}}\rangle = |r_{\text{DW}}|^2 \left(M + 2\Re\left\{ \sum_{p=1}^{M-1} \sum_{j=1}^{M-p} \left\langle e^{\imath \Delta\phi_{p,j}} \right\rangle \right\} \right) \tag{A.14}$$

where $\Re\{\cdot\}$ is the real part and $\Delta\phi_{p,j} = k(\Delta z_p - \Delta z_{p+j})$ is the phase difference. From equation (A.11) follows that

$$\Delta\phi_{p,j} = jk\Lambda_{\text{QPM}} + 2k(\delta_{\text{DE},p} - \delta_{\text{DE},j}) + 2k \sum_{q=p+1}^{p+j} \delta_{\text{PE},q}. \tag{A.15}$$

The position deviations δ_{DE} and δ_{PE} are assumed to be statistically independent and with a normal distribution function with standard deviations σ_{DE} and σ_{PE} and zero mean.

Since equation (A.15) is a linear combination of the deviations, the phase difference $\Delta\phi_{p,j}$ can be described by a joint normal probability density function $P_{\Delta\phi}$:

$$P_{\Delta\phi}(\Delta\phi_{p,j}) = \frac{1}{\sqrt{2\pi}\sigma_{\Delta\phi_{p,j}}} \exp\left(-\frac{(\Delta\phi_{p,j} - \langle\Delta\phi_{p,j}\rangle)^2}{2\sigma^2_{\Delta\phi_{p,j}}}\right). \tag{A.16}$$

The mean $\langle\Delta\phi_{p,j}\rangle$ and standard deviation $\sigma_{\phi_{p,j}}$ can be obtained directly from equation (A.15), which leads to

$$\langle\Delta\phi_{p,j}\rangle = jk\Lambda_{\mathrm{QPM}} \tag{A.17}$$

and

$$\sigma^2_{\phi_{p,j}} = 8k^2\sigma^2_{\mathrm{DE}} + 4jk^2\sigma^2_{\mathrm{PE}}. \tag{A.18}$$

With the probability density function (A.16), the ensemble average in equation (A.14) can be calculated by

$$\left\langle e^{\imath\Delta\phi_{p,j}}\right\rangle = \int P_{\Delta\phi}(\Delta\phi_{p,j}) e^{\imath\Delta\phi_{p,j}} \mathrm{d}\Delta\phi_{p,j} \tag{A.19}$$

that can be solved analytically:

$$\left\langle e^{\imath\Delta\phi_{p,j}}\right\rangle = e^{-\frac{1}{2}\sigma^2_{\Delta\phi_{p,j}} + \imath\langle\Delta\phi_{p,j}\rangle}. \tag{A.20}$$

Inserting equations (A.17), (A.18) and (A.20) into equation (A.14) gives

$$\langle R_{\mathrm{PP,\ plane}}\rangle = |r_{\mathrm{DW}}|^2 \left(M + 2\Re\left\{\sum_{p=1}^{M-1}\sum_{j=1}^{M-p} e^{-4k^2\sigma^2_{\mathrm{DE}} - 2jk^2\sigma^2_{\mathrm{PE}} + \imath jk\Lambda_{\mathrm{QPM}}}\right\}\right). \tag{A.21}$$

After splitting off a factor of $\exp(-4k^2\sigma^2_{\mathrm{DE}})$, the inner sum can be calculated as a finite geometric series:

$$\langle R_{\mathrm{PP,\ plane}}\rangle = |r_{\mathrm{DW}}|^2 \left(M + 2e^{-4k^2\sigma^2_{\mathrm{DE}}}\Re\left\{\sum_{p=1}^{M-1} \frac{1 - e^{-(M-p)(2k^2\sigma^2_{\mathrm{PE}} - \imath k\Lambda_{\mathrm{QPM}})}}{e^{2k^2\sigma^2_{\mathrm{PE}} - \imath k\Lambda_{\mathrm{QPM}}} - 1}\right\}\right). \tag{A.22}$$

Again, the sum is a finite geometric series that is calculated to obtain the final result

$$\begin{aligned}\langle R_{\mathrm{PP,\ plane}}\rangle &= |r_{\mathrm{DW}}|^2 \\ &\times \left(M + 2e^{-4k^2\sigma^2_{\mathrm{DE}}}\Re\left\{\frac{M-1}{e^{2k^2\sigma^2_{\mathrm{PE}} - \imath k\Lambda_{\mathrm{QPM}}} - 1} + \frac{e^{-(M-1)(2k^2\sigma^2_{\mathrm{PE}} - \imath k\Lambda_{\mathrm{QPM}})} - 1}{\left(e^{2k^2\sigma^2_{\mathrm{PE}} - \imath k\Lambda_{\mathrm{QPM}}} - 1\right)^2}\right\}\right).\end{aligned} \tag{A.23}$$

A.2.2. Nonlinear Conversion Efficiency

The ensemble average of the nonlinear conversion efficiency $\langle\eta\rangle$ for second harmonic generation is defined as [30]:

$$\langle\eta\rangle = \frac{1}{M^2}\left\langle\left|\sum_{p=1}^{M} e^{-\imath\Delta k_{\mathrm{QPM}}\delta z_p}\right|^2\right\rangle. \tag{A.24}$$

M is the total number of domains and $\Delta k_{\mathrm{QPM}} = 2\pi/\Lambda_{\mathrm{QPM}}$ is the quasi phase matching wave vector with the poling period Λ_{QPM}. The deviation δz_p of the pth domain wall (DW) is given by

$$\delta z_p = \delta_{\mathrm{DE},p} + \sum_{q=1}^{p} \delta_{\mathrm{PE},q} \tag{A.25}$$

where δ_{DE} and δ_{PE} are the deviations of the DW position according to duty cycle and period error, respectively.

Similar to the derivation for the reflectance, the DW position deviations are described by a joint normal distribution. The corresponding average conversion efficiency is then

$$\langle\eta\rangle = \frac{1}{M} + \frac{2}{M^2}e^{-\Delta k_{\mathrm{QPM}}^2\sigma_{\mathrm{DE}}^2}\left[\frac{e^{\frac{1}{2}\Delta k_{\mathrm{QPM}}^2\sigma_{\mathrm{PE}}^2}\left(M + e^{-\frac{1}{2}M\Delta k_{\mathrm{QPM}}^2\sigma_{\mathrm{PE}}^2} - 1\right) - M}{\left(e^{\frac{1}{2}\Delta k_{\mathrm{QPM}}^2\sigma_{\mathrm{PE}}^2} - 1\right)^2}\right]. \tag{A.26}$$

σ_{DE} and σ_{PE} are the standard deviations of the DW position errors for the duty cycle error and the period error, respectively.

A.3. Sellmeier Equation

The temperature and wavelength dependence of the refractive index of periodically poled nonlinear crystals is modeled by a Sellmeier equation. Combining the results from [148], [149], the equation for both lithium niobate and lithium tantalate crystals, has the form

$$\begin{aligned} n(\lambda, T) = \sqrt{a_1 + \frac{a_2}{\lambda^2 - a_3} + \frac{a_4}{\lambda^2 - a_5} + \frac{a_6}{\lambda^2 - a_7} + a_8\lambda^2} \\ + \left(\frac{a_9}{\lambda^4} + \frac{a_{10}}{\lambda^3} + \frac{a_{11}}{\lambda^2} + \frac{a_{12}}{\lambda} + a_{13} + a_{14}\lambda\right)\left(a_{15}(T - T_a) + a_{16}(T - T_a)^2\right), \end{aligned} \tag{A.27}$$

where λ and T are the wavelength and the temperature, respectively. T_a is the reference temperature and a_1 to a_{16} are material specific model parameters.

A.4. Thermal Expansion

The thermal expansion of periodically poled nonlinear crystals is described by a linear thermal expansion model. The actual length $l(T)$ at the temperature T can be calculated by

$$l(T) = l(T_0)[1 + \alpha_{\mathrm{th}}(T - T_0)]. \tag{A.28}$$

$l(T_0)$ is the given length at the reference temperature T_0 and α_{th} is the coefficient of linear thermal expansion.

A.5. Guided Modes in Channel Waveguide Periodically Poled Nonlinear Crystals

To calculate the field distribution and related properties of guided optical modes in channel waveguide periodically poled nonlinear crystals (PPNC), different approaches are used for the ridge waveguide (RW) and depressed cladding waveguide (DCW) configurations. In section 3.2.2 details about the investigated channel waveguide PPNCs are given. The RW PPNCs have design wavelengths of 1178 nm and 1122 nm, while the DCW PPNC has a design wavelength of 1064 nm.

A.5.1. Ridge Waveguide Periodically Poled Nonlinear Crystals

The RW PPNC is characterized by strong optical confinement and high refractive index contrast between the waveguide core and cladding. In figure A.1a the simplified geometry of the cross section of the waveguide region of the RW PPNC is shown. In the vertical

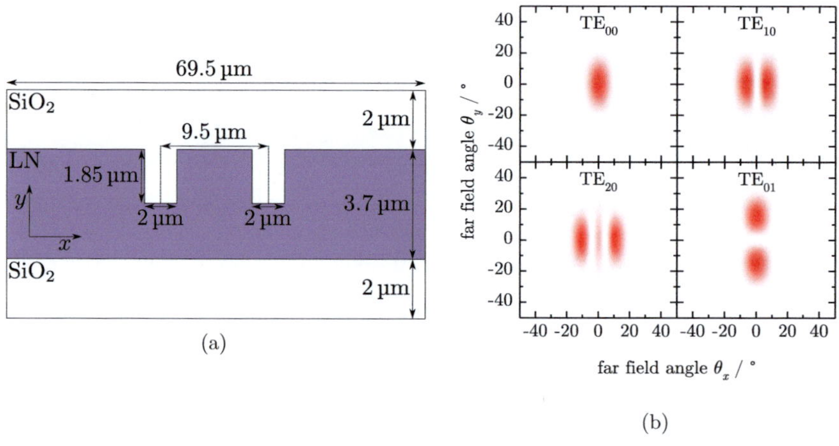

Figure A.1.: (a) Simplified geometry of the ridge waveguide (RW) periodically poled nonlinear crystal. The planar layer of lithium niobate (LN) is surrounded by the silicon dioxide (SiO_2) cladding while the RW is formed in the center by two trenches of SiO_2. (b) Far field intensity distributions of four guided modes TE_ν at a wavelength of 1178 nm. The calculations are based on the electric field distributions obtained with the software *FIMMWAVE*.

y-direction, the light is guided in a 3.7 µm thick planar layer of lithium niobate (LN), surrounded by a silicon dioxide (SiO_2) cladding with a thickness of 2 µm. The RW is defined in the horizontal x-direction by two 2 µm wide rectangular SiO_2 trenches which have a distance of 9.5 µm and reach 1.85 µm into the LN layer. The refractive indices of LN and SiO_2 are assumed to be 2.145 and 1.45, respectively, while dispersion is neglected.

Due to the high refractive index contrast and the irregular geometry of the waveguide, the guided modes have to be determined by numerical methods. The calculation is done with the commercial software *FIMMWAVE*. Since the incident light in the RW PPNCs is polarized along the x-direction, only TE modes are considered. In addition to the fundamental TE_{00} mode several higher-order guided modes are possible.

The far field intensity distributions at a wavelength of 1178 nm of the fundamental mode and three higher-order modes with the highest propagation constant β_ν are shown in figure A.1b. While the electric field distributions are obtained from *FIMMWAVE*, the far field intensity distributions are calculated by Fourier transform (equation (3.5) in section 3.1). The modes TE_{10} and TE_{20} correspond to the horizontal higher-order modes whereas the mode TE_{01} is the first vertical higher-order mode. The corresponding effective index differences $\Delta n_{\mathrm{eff},\nu}$ are calculated from the propagation constants by

$$\Delta n_{\mathrm{eff},\nu} = \frac{\beta_\nu \lambda}{2\pi} - n_{\mathrm{LN}}, \tag{A.29}$$

where λ is the wavelength and n_{LN} is the refractive index of LN. The results are given in table A.1.

Table A.1.: Calculated effective index difference $\Delta n_{\mathrm{eff},\nu}$ for guided TE_ν modes of the ridge waveguide at a wavelength of 1178 nm.

mode TE_ν	TE_{00}	TE_{10}	TE_{20}	TE_{01}
effective index difference $\Delta n_{\mathrm{eff},\nu}/10^{-3}$	−6.5	−10.2	−15.8	−22.3

At a wavelength of 1122 nm almost the same results are obtained, but the effective index differences are slightly smaller. For the fundamental TE_{00} mode an effective index difference of -5.9×10^{-3} is determined.

A.5.2. Depressed Cladding Waveguide Periodically Poled Nonlinear Crystals

The DCW PPNC consists of a radially symmetric and weakly guiding waveguide. Figure A.2a shows the simplified geometry of the core region. The lithium tantalate (LT) waveguide core has a diameter of 36 µm which is surrounded by an infinitely thick index depressed cladding (IDC). According to the Sellmeier equation (A.27) and the parameters in table B.2, a refractive index of 2.134 is used for the waveguide core. While the actual refractive index of the IDC is not available, it can be assumed to be smaller by 1×10^{-2} [150], yielding a refractive index of 2.124.

For the radial waveguide with only a small refractive index contrast and linearly polarized incident light, LP modes can be calculated directly [151]. In polar coordinates,

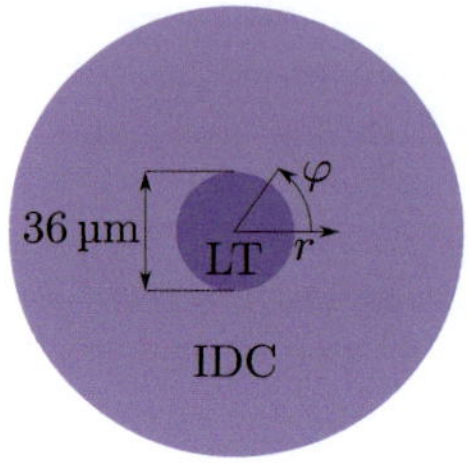

(a)

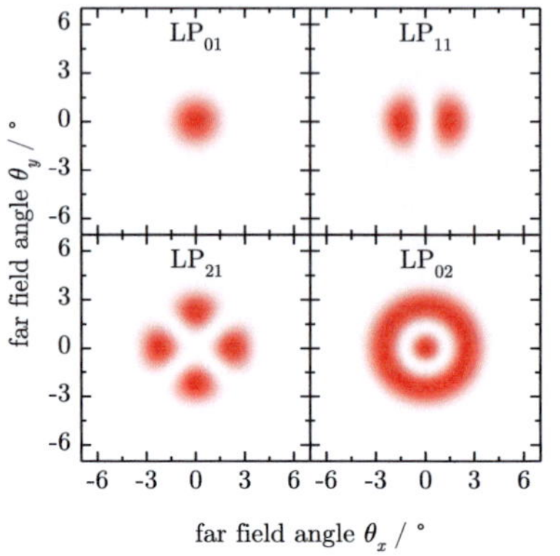

(b)

Figure A.2.: (a) Simplified geometry of the depressed cladding waveguide periodically poled nonlinear crystal. The circular lithium tantalate (LT) core is surrounded by the index depressed cladding (IDC). (b) Far field intensity distributions of four guided modes LP_ν at a wavelength of 1064 nm. The calculations are based on the electric field distributions obtained with analytical expressions using Bessel functions.

the electric field distribution E_{lm} of the LP mode is given by

$$E_{lm}(r,\varphi) = \begin{cases} \frac{E_0}{J_l(u_{lm})} J_l\left(\frac{u_{lm} r}{r_{\text{core}}}\right)\cos(l\varphi) & r < r_{\text{core}} \\ \frac{E_0}{K_l(w_{lm})} K_l\left(\frac{w_{lm} r}{r_{\text{core}}}\right)\cos(l\varphi) & r \geq r_{\text{core}} \end{cases}, \tag{A.30}$$

where r_{core} is the core radius, half of the core diameter. J_l is the Bessel function of the first kind and K_l is the modified Bessel function of the second kind. The azimuthal index l and the radial index m are integers numbering the LP mode with $l \in \mathbb{N}_0$ and $m \in \mathbb{N}$. The parameters u_{lm} and w_{lm} are defined by

$$u_{lm} = r_{\text{core}}\sqrt{n_{\text{core}}^2 k_0^2 - \beta_{lm}^2} \tag{A.31}$$

$$w_{lm} = r_{\text{core}}\sqrt{\beta_{lm}^2 - n_{\text{clad}}^2 k_0^2}. \tag{A.32}$$

Here, n_{core} and n_{clad} are the refractive indices of the core and cladding, respectively. $k_0 = 2\pi/\lambda$ is the vacuum wave vector and β_{lm} is the propagation constant. The propagation constants of guided modes can be determined as the solutions to

$$\frac{u_{lm} J_{l+1}(u_{lm})}{J_l(u_{lm})} = \frac{w_{lm} K_{l+1}(w_{lm})}{K_l(w_{lm})}. \tag{A.33}$$

which follows from the continuity requirement of the electric field and its spatial derivative. For each azimuthal index l, the radial index m numbers the modes in descending order with respect to the propagation constant.

Similar to the RW, several higher-order guided modes are possible for the DCW. In figure A.2b the calculated far field intensity distributions for the fundamental LP_{01} and the three higher order modes with the largest propagation constant are shown. While the electric field distribution is obtained from equation A.30, the far field intensity distributions are again calculated by equation (3.5) after transformation to Cartesian coordinates. The corresponding effective index differences are determined from the propagation constants by equation (A.29) and are given in table A.2. Due to the large core

Table A.2.: Calculated effective index difference $\Delta n_{\text{eff},\nu}$ for guided LP_ν modes of the depressed cladding waveguide at a wavelength of 1064 nm.

mode LP_ν	LP_{01}	LP_{11}	LP_{21}	LP_{02}
effective index difference $\Delta n_{\text{eff},\nu}/10^{-4}$	−1.1	−2.8	−5.0	−5.8

diameter and the low refractive index contrast between core and cladding, the effective index differences are of the order of only -1×10^{-4}.

A.6. Computation Scheme of the Multiphysics Model

The multiphysics model described in section 4.1.1 consists of three models for the optical field, the carriers and the temperature distribution. Since these are mutually coupled, an appropriate computational scheme is required to resolve the respective dependencies. Each model is based on a differential equation, that is solved by an individual numerical method, hereafter referred to as the optical, electrical and thermal solvers. All solvers operate on the same spatial grid, which discretizing the horizontal-longitudinal $(x - z)$ plane.

The paraxial wave equation (4.9) is solved by a fast Fourier transform split-step approach as used in [90]. For the nonlinear diffusion equation (4.15) a deferred correction technique in combination with Newton's method similar to [74] is used. Thereby the diffusion equation is transformed into a system of first-order differential equations and integrated using the trapezoidal rule. The solution of the heat equation (4.22) is obtained by a combination of Green's function approach and Fourier transformation that is specifically optimized for the typical planar layer geometry of the SOA devices. A detailed description of the thermal solver is given in section A.6.1.

The coupling of the three solvers in combination with the boundary conditions is visualized in the flowchart in figure A.3. The algorithm successively calculates the field amplitudes of the forward and backward propagating light E^+ and E^-. This procedure is iterated until the results converge to the desired degree [152].

For each spatial grid point z_i the excess carrier distribution N is calculated from the temperature distribution T and the field amplitude of the forward propagating wave E^+

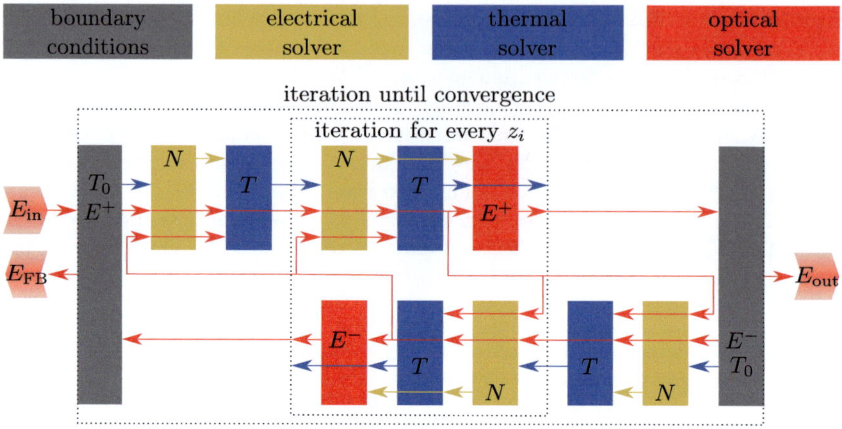

Figure A.3.: Flowchart of the algorithm for the simulation of light propagating in z-direction through semiconductor optical amplifiers (SOA). The application of boundary conditions or solvers to the physical models are symbolized by solid blocks and provide the field amplitudes of the forward and backward propagating light E^+ and E^-, the excess carrier distribution N and the temperature distribution T. The incident field amplitude is E_{in}, while E_{out} and E_{FB} are the amplitudes of the output and feedback emission. T_0 is the reference temperature.

of the previous grid point. The field amplitude of the backward propagating wave E^- is taken from the next grid point. Since the amplitude of the backward propagating wave is undefined during the first iteration, E^- is set to zero. From the excess carrier distribution and the field amplitudes of the previous grid point, the temperature distribution T for the current grid point is obtained. Due to the iteration order, the temperature at the first grid point is undefined. Therefore, the temperature is obtained from a preliminary excess carrier distribution that is calculated for the constant reference temperature T_0. Finally, the new electric field amplitude E^+ is calculated. The field amplitude of the forward propagating wave E^+ of the first grid point (at the back facet) is defined by the boundary condition where the incident field amplitude E_{in} is required.

At the last grid point (at the front facet) the boundary condition defines the amplitude E_{out} of the forward emitted light. For the amplitude of the backward propagating wave E^- the same iteration order is applied, using the forward propagating amplitudes E^+ of the previous iteration. With the iteration back to the first grid point, the amplitude of the feedback light E_{FB} is given.

A.6.1. Thermal Solver

The thermal model and the tailored computational algorithm presented in this section have been published with slight modifications in [153].

In order to solve the two-dimensional heat equation

$$[\partial_x \kappa(x,y)\partial_x + \partial_y \kappa(x,y)\partial_y]\, T(x,y) = -H(x,y), \tag{A.34}$$

the problem is first modified with respect to the horizontal (x) Neumann boundary conditions and the special geometry. The semiconductor optical amplifier (SOA) can be considered as a vertical stack of different layers with constant width, possibly including heat spreaders. Horizontal structures such as etched trenches are much smaller than the device thickness and are therefore neglected. Thus, the thermal conductivity is assumed to be constant in the horizontal direction and piecewise constant in the vertical (y) direction. The problem can be further simplified by considering the general volume heat flux $H(x,y)$ in detail. The major part of the heat is generated in the active zone or in the thin p-doped layers above it, which are much smaller compared to the device dimensions. Therefore, the volume heat flux can be assumed to be a sheet heat flux q, located at the active zone at $y = s$:

$$H(x,y) \approx q(x)\delta(y-s). \tag{A.35}$$

The sheet heat flux is given by

$$q(x) = \int H(x,y)\mathrm{d}y. \tag{A.36}$$

The temperature distribution T with respect to the reference temperature T_0 and the sheet heat flux q can be expanded in terms of cosine series (which is equivalent to a cosine transformation of the problem), satisfying the Neumann boundary conditions

$$T(x,y) - T_0 = \frac{2}{w}\sum_{m=0}^{\infty}{}' t(k_m, y)\cos(k_m x) \tag{A.37}$$

$$q(x) = \frac{2}{w}\sum_{m=0}^{\infty}{}' \theta(k_m)\cos(k_m x). \tag{A.38}$$

Here, $k_m = \frac{m\pi}{w}$ and the coefficients are given by

$$t(0,y) = \int_0^w T(x,y)\mathrm{d}x \tag{A.39}$$

$$t(k_m,y) = \int_0^w T(x,y)\cos(k_m x)\mathrm{d}x \tag{A.40}$$

and similarly for θ. The prime notation at the sum symbol means that the terms for $k_m = 0$ must be multiplied by $1/2$.

The heat equation is then

$$\left[\partial_y \kappa(y)\partial_y - \kappa(y)k_m^2\right] t(k_m,y) = -\theta(k_m)\delta(y-s). \tag{A.41}$$

The inhomogeneous ordinary differential equation (A.41) can be solved by the Green's function method. The Green's function g is a solution of the equation

$$\left[\partial_y \kappa(y) \partial_y - \kappa(y) k_m^2\right] g(k_m, y, y') = \delta(y - y'), \tag{A.42}$$

subject to the same boundary conditions as T. With the solution of equation (A.42) the solution of the inhomogeneous problem (A.41) can be calculated by integration:

$$t(k_m, y) = -\int g(k_m, y, y')\theta(k_m)\delta(y' - s)\mathrm{d}y' = -g(k_m, y, s)\theta(k_m), \tag{A.43}$$

where equation (A.35) is used. Therefore, equation (A.42) has to be solved only for $y' = s$. This leads to the final solution for the temperature

$$T(x, y) = T_0 - \frac{2}{L}\sum_{m=0}^{\infty}{}' g(k_m, y, s)\theta(k_m)\cos(k_m x). \tag{A.44}$$

The final result is ideally suited for the efficient calculation of the thermally induced effective index change given by equation (4.12). Inserting (A.44) into (4.12) yields

$$\Delta n_T(x) = -n_T' \frac{2}{L}\sum_{m=0}^{\infty}{}' K(k_m)\theta(k_m)\cos(k_m x) \tag{A.45}$$

with

$$K(k_m) = \int g(k_m, y, s)\Phi^2(y)\mathrm{d}y. \tag{A.46}$$

The kernel K depends only on the SOA geometry and boundary conditions but not on the horizontal heat source distribution. Therefore, the integration along y does not necessarily need to be recalculated for each sheet heat flux distribution q or θ. The backward transformation from k_m based quantities of the cosine series to real space coordinates in equation (A.45) can be efficiently performed by the discrete cosine transformation (DCT). Thus, the thermally induced refractive index change is calculated from the heat source distribution using only two DCTs. First, a DCT is applied to the sheet heat flux q, then the result is multiplied by the pre-computed kernel and finally the inverse DCT is applied to obtain the final result.

Calculation of Green's Function

The general solution of (A.42) for $y \neq s$ within a layer $i = 1, \cdots, N$ with $y_i \leq y < y_{i+1}$, i.e. thickness $d_i = y_{i+1} - y_i$, can be written as

$$g_i(k_m, y, s) = \begin{cases} A_i(k_m, s)e^{k_m(y-s)} + B_i(k_m, s)e^{2k_m y_i}e^{-k_m(y+s)} & m \neq 0, y < s \\ C_i(k_m, s)e^{-k(y-s)} + D_i(k_m, s)e^{-2k_m y_{i+1}}e^{k(y+s)} & m \neq 0, y > s \\ A_i(k_m, s)(y - y_i) + B_i(k_m, s) & m = 0, y < s \\ C_i(k_m, s)(y_{i+1} - y) + D_i(k_m, s) & m = 0, y > s. \end{cases} \tag{A.47}$$

Here, $y = s$ is treated as a layer interface, even though the material below and above can be the same. The derivative is given by

$$\partial_y g_i(k_m, y, s) = \begin{cases} A_i(s)k_m e^{k_m(y-s)} - B_i(s)k_m e^{2k_m y_i} e^{-k_m(y+s)} & m \neq 0, y < s \\ -C_i(s)k_m e^{-k_m(y-s)} + D_i(s)k_m e^{-2k_m y_{i+1}} e^{k_m(y+s)} & m \neq 0, y > s \\ A_i(k_m, s) & m = 0, y < s \\ -C_i(k_m, s) & m = 0, y > s. \end{cases} \tag{A.48}$$

The coefficients A_i, B_i, C_i and D_i depending on the position s of the heat source, are determined from the boundary and continuity conditions. The mixed boundary conditions at the bottom ($y = y_1$) and top ($y = y_{N+1}$) boundary read as follows

$$\partial_y g_1(y_1) - \sigma_\mathrm{b} g_1(y_1) = 0 \tag{A.49}$$

$$\partial_y g_N(y_{N+1}) + \sigma_\mathrm{t} g_N(y_{N+1}) = 0, \tag{A.50}$$

while $\sigma_\mathrm{b,t} = 0$ corresponds to a pure Neumann condition and $\sigma_\mathrm{b,t} \to \infty$ corresponds to a pure Dirichlet condition.

At the hetero boundaries $y = y_i$ continuity conditions for the Green's function

$$g_i(y_{i+1}) - g_{i+1}(y_{i+1}) = 0 \tag{A.51}$$

and its normal derivative associated with the heat flux

$$\kappa_i \partial_y g_i(y_{i+1}) - \kappa_{i+1} \partial_y g_{i+1}(y_{i+1}) = 0 \quad y_{i+1} \neq s \tag{A.52}$$

$$\kappa_i \partial_y g_i(s) - \kappa_{i+1} \partial_y g_{i+1}(s) = -1 \quad y_{i+1} = s \tag{A.53}$$

apply.

Equations (A.49) to (A.53) yield a total number of $2N$ relations for the $2N$ unknown parameters A_i, B_i, C_i and D_i of the N layers.

Algorithm

Starting with the first and last layers, the initial conditions are set to $B_1 = 1$ (σ_b small) or $B_1 = 1/\sigma_\mathrm{b}$ (σ_b large) and $D_N = 1$ (σ_t small) or $D_N = 1/\sigma_\mathrm{t}$ (σ_t large). The coefficients in the adjacent layers are then determined using the relations (A.51) and (A.52) and so on until the layers directly below and above the heat source layer (i with $y_i < s$ and $i+1$ with $y_{i+2} > s$) are reached. To fulfill the conditions (A.51) and (A.53) B_1 and D_N have to be scaled by factors f_1 and f_2. These factors are determined by rewriting equations (A.51) and (A.53) as

$$\begin{pmatrix} g_i(s) & -g_{i+1}(s) \\ \kappa_i \partial_y g_i(s) & -\kappa_{i+1} \partial_y g_{i+1}(s) \end{pmatrix} \begin{pmatrix} f_1 \\ f_2 \end{pmatrix} = \begin{pmatrix} 0 \\ -1 \end{pmatrix}. \tag{A.54}$$

The solution of (A.54) is

$$\begin{pmatrix} f_1 \\ f_2 \end{pmatrix} = \frac{1}{g_i(s)\kappa_{i+1}\partial_y g_{i+1}(s) - g_{i+1}(s)\kappa_i \partial_y g_i(s)} \begin{pmatrix} g_{i+1}(s) \\ g_i(s) \end{pmatrix}. \tag{A.55}$$

To obtain the final solution the coefficients (A_i, B_i) and (C_i, D_i) have to be multiplied by f_1 and f_2, respectively.

The main advantage of this approach is that instead of discretizing the entire device, only the vertical interfaces need to be considered. Furthermore, the vertical sampling of the temperature distribution can be chosen arbitrarily without loss of accuracy.

Thermal Model Validation

In order to validate the thermal solver, the temperature distribution in a simple exemplary geometry is calculated and compared with the results of a commercial finite element method (FEM) software (*ANSYS 2021*).

The geometry is shown in figure A.4a. The sample consists of a 800 µm wide gallium

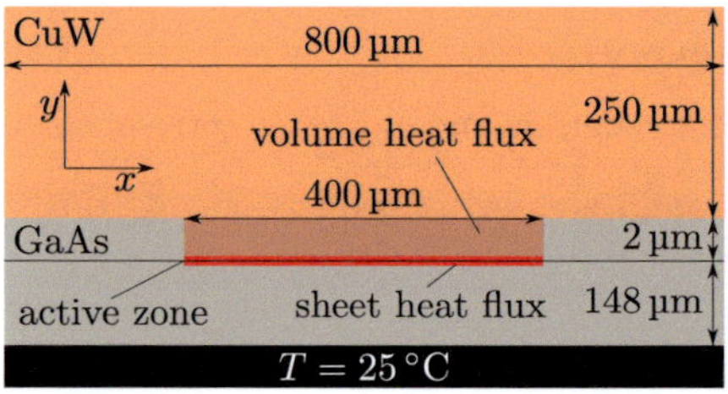

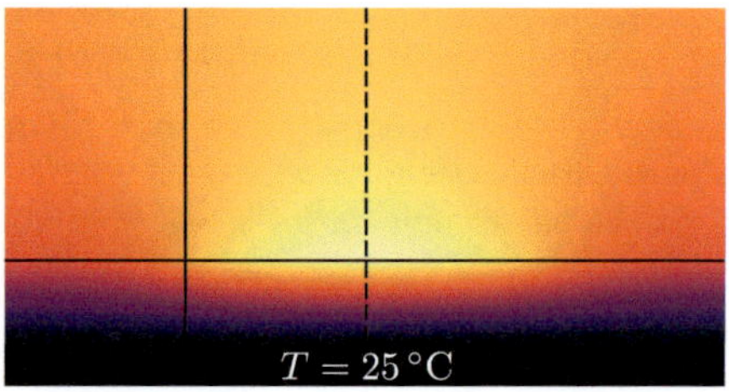

Figure A.4.: (a) Sample geometry used for the validation of the thermal solver. A gallium arsenide (GaAs) chip is placed between a heat sink (black) with a temperature of $T = 25\,°C$ and a copper-tungsten (CuW) heat spreader. The red line or red shaded area symbolize the locations of the sheet heat flux and volume heat flux, respectively. (b) Corresponding temperature distribution in the sample geometry. The horizontal and vertical solid or dashed lines mark the locations at which temperature profiles are obtained for further evaluations.

arsenide (GaAs) chip that is mounted on a heat sink with $T = 25\,°C$ (Dirichlet condition). The chip is divided into two sections with a thickness of 148 µm and 2 µm, where the interface is assumed to be the active zone. A 250 µm thick copper-tungsten (CuW) heat spreader is attached to the top of the chip. This geometry is typical for e.g. the taper section of tapered amplifiers (see section 4.2.2). The thermal conductivities used for GaAs and CuW are $55\,W\,m^{-1}\,K^{-1}$ and $180\,W\,m^{-1}\,K^{-1}$, respectively. Homogeneous Neumann conditions apply at all boundaries except at the bottom. Heat is generated in a horizontally centered strip of 400 µm width. There are two different cases to be distinguished here. First, the heat source is modeled by a sheet heat flux of $5\,W/0.4\,mm^2$, vertically centered in the active zone, which is consistent with the assumptions made for the Green's function approach. In figure A.4a this is symbolized by the red line in the active zone. Second, the heat source is modeled by a spatially extended volume heat flux, that extends from the active zone to the interface between GaAs and CuW. This 2 µm thick heat source is marked in figure A.4a by the red shaded rectangle. The applied volume heat flux is $5\,\mathrm{W}/(0.4\,\mathrm{mm}^2 \cdot 2\,\mu\mathrm{m})$. The FEM software is used to solve both geometries with sheet and volume heat fluxes. The results are compared with the solution of the Green's function approach that naturally assumes the geometry with sheet heat flux.

Figure A.4b shows the heat map of the temperature distribution in the sample geometry obtained by the Green's function method. The horizontal and vertical lines mark the positions where temperature profiles are taken in order to compare the three simulation results. The horizontal temperature profile is taken at the active zone at $y = 148\,\mu\mathrm{m}$ and two vertical profiles are evaluated at $x = -200\,\mu\mathrm{m}$ and $x = 0\,\mu\mathrm{m}$ (vertical solid and dashed lines). The temperature profiles are shown in figure A.5. Both the Green's

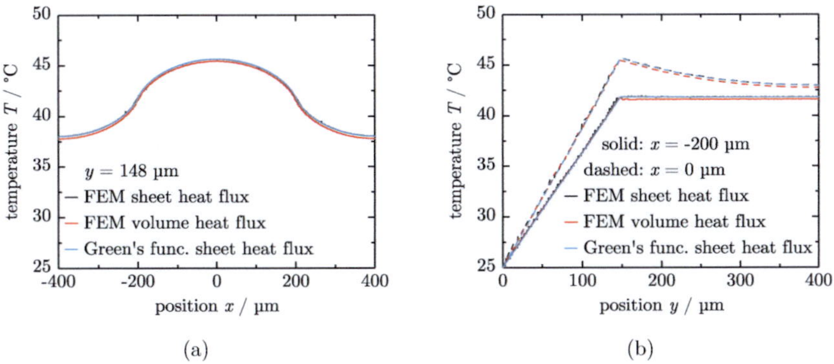

Figure A.5.: Comparison of temperature profiles obtained by finite element method (FEM) and Green's function method at different positions of the sample geometry. (a) Horizontal temperature distributions at the vertical position $y = 148\,\mu\mathrm{m}$ of the active zone. (b) Vertical temperature distributions at two horizontal positions $x = -200\,\mu\mathrm{m}$ and $x = 0\,\mu\mathrm{m}$.

function method and the FEM approach give almost the same results for the geometry with a sheet heat flux. Small fluctuations in the temperature obtained by the FEM software are post processing artifacts. Compared to the temperature of the sheet heat flux geometry, the volume heat source leads to the same temperature distribution but an almost constant offset of $-0.2\,\mathrm{K}$. Especially when considering the effect on the refractive index, this offset is negligible and the sheet heat source approximation is well suited.

The model based on the Green's function method is only applicable to rectangular geometries which may differ from real-word assemblies. Especially for the epi-side up mounting scheme, the heat spreader mounted to the top can be wider than the actual chip underneath. In figure A.6a a sample geometry with a wide heat spreader is shown. The geometry is similar to that shown in figure A.4a except for the wide heat spreader

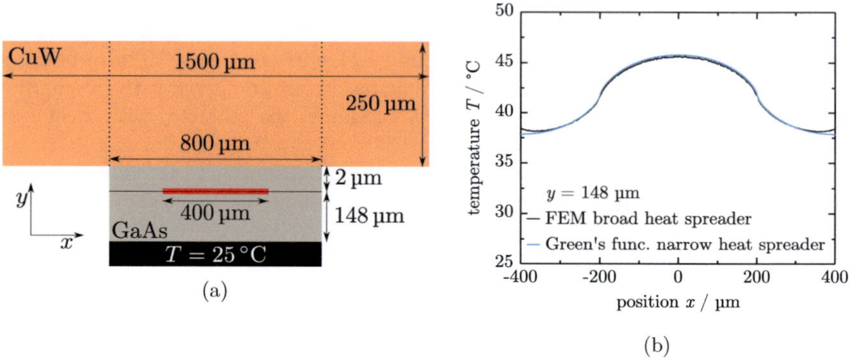

Figure A.6.: (a) Sample geometry of a gallium arsenide (GaAs) chip placed between a heat sink (black) with a temperature of $T = 25\,°\mathrm{C}$ and a wide copper-tungsten (CuW) heat spreader. The red line symbolizes the location of the sheet heat flux. With the vertical black dotted lines the rectangular geometry used for the Green's function method is marked. (b) Comparison of horizontal temperature profiles obtained by finite element method (FEM) and Green's function method at the vertical position $y = 148\,\mu\mathrm{m}$ of the active zone.

with a width of 1500 µm. The areas of the heat spreader that extend beyond the chip are completely surrounded by boundaries with homogeneous Neumann conditions. Thus, the net heat flow into these areas vanishes and their contribution to the heat dissipation is expected to be small.

In figure A.6b the horizontal temperature profiles for the geometry with wide heat spreader and the rectangular geometry are shown. The profiles are again obtained at the vertical position of the active zone ($y = 148\,\mu\mathrm{m}$) and the heat source is modeled as a sheet heat flux of $5\,\mathrm{W}/0.4\,\mathrm{mm}^2$. While the FEM software is used to calculate the temperature for the geometry with wide heat spreader, the temperature for the rectangular geometry is obtained from the Green's function method, which is identical to the data shown in figure A.5a. Both temperature profiles are almost similar. However, the temperatures for the geometry with wide heat spreader are lower by 0.15 K in a wide region around the center. In contrast, the temperature close to the boundaries is increased by up to 0.55 K. Since the deviations are small compared to the overall temperature variations and the largest deviations occur near the device boundary, the effects on the propagating optical waves can be neglected. Thus, for epi-side up mounting schemes with wide heat spreaders, the temperature can be calculated based on a rectangular geometry and the Green's function method can still be used.

A.7. Calculation of the Position of the Beam Waist

The position of the beam waist along the beam direction of a light beam propagating in the z-direction can be calculated using the Wigner distribution function (WDF) and its transformation properties [116]–[118]. In one dimension, the Wigner distribution function is defined as

$$h(x,u) = \frac{\varepsilon_0 c_0 k_0}{4\pi} \int E^*\left(x+\frac{s}{2},z\right) E\left(x-\frac{s}{2},z\right) e^{\imath k_0 u s} \mathrm{d}s, \tag{A.56}$$

where E is the electric field amplitude of the light beam at an arbitrary position z. ε_0, c_0 and k_0 are the vacuum values of the permittivity, speed of light, and wave vector, respectively. * denotes the complex conjugate. The WDF is normalized such that the optical power P is given by $P = \int h(x,u)\mathrm{d}x\mathrm{d}u$. The moments of the WDF are defined by

$$\langle x^m u^n \rangle = \frac{1}{P} \iint h(x,u) x^m u^n \mathrm{d}x\mathrm{d}u \tag{A.57}$$

with $m, n \in \mathbb{N}_0$. To describe the beam properties, the centralized second-order moments

$$\langle x^m u^n \rangle_\mathrm{c} = \langle x^m u^n \rangle - \langle x \rangle^m \langle u \rangle^n \tag{A.58}$$

with $m+n=2$ are of particular interest. From the transformation properties of the moments, a relation can be derived for the beam diameter d as a function of the longitudinal coordinate z':

$$d^2(z') \propto \langle x^2 \rangle_\mathrm{c} + 2\langle xu \rangle_\mathrm{c} z' + \langle u^2 \rangle_\mathrm{c} z'^2. \tag{A.59}$$

The coordinate z' is the relative position with respect to the position of the initial electric field distribution. The beam waist position z_BW is then found as the position of the minimum of d:

$$z_\mathrm{BW} = -\frac{\langle xu \rangle_\mathrm{c}}{\langle u^2 \rangle_\mathrm{c}}. \tag{A.60}$$

Thus, the moments $\langle x \rangle$, $\langle u \rangle$, $\langle xu \rangle$ and $\langle u^2 \rangle$ have to be calculated. For the sake of clarity, the z-dependence of E is omitted in the following.

- The first-order moment regarding x is defined as:

  $$\langle x \rangle = \frac{\varepsilon_0 c_0 k_0}{4\pi P} \iiint E^*\left(x+\frac{s}{2}\right) E\left(x-\frac{s}{2}\right) e^{\imath k_0 u s} x \mathrm{d}s\mathrm{d}x\mathrm{d}u. \tag{A.61}$$

  With the identity $\delta(s') = \int \exp(\imath 2\pi u s')\mathrm{d}u$, where $\delta(\cdot)$ is the delta distribution, follows that

  $$\langle x \rangle = \frac{\varepsilon_0 c_0 k_0}{4\pi P} \iint E^*\left(x+\frac{s}{2}\right) E\left(x-\frac{s}{2}\right) \delta\left(\frac{k_0 s}{2\pi}\right) x \mathrm{d}s\mathrm{d}x. \tag{A.62}$$

  Using the scaling properties of the delta distribution leads finally to

  $$\langle x \rangle = \frac{\varepsilon_0 c_0}{2P} \int |E(x)|^2 x \mathrm{d}x. \tag{A.63}$$

- The first-order moment regarding u is defined as

$$\langle u\rangle = \frac{\varepsilon_0 c_0 k_0}{4\pi P}\iiint E^*\left(x+\frac{s}{2}\right)E\left(x-\frac{s}{2}\right)e^{\imath k_0 us}u\mathrm{d}s\mathrm{d}x\mathrm{d}u. \tag{A.64}$$

Using partial integration and the fact that $E(x)$ vanishes for $x\to\infty$ gives

$$\langle u\rangle = -\frac{\varepsilon_0 c_0}{4\pi P}\iiint \frac{1}{\imath}\partial_s\left[E^*\left(x+\frac{s}{2}\right)E\left(x-\frac{s}{2}\right)\right]e^{\imath k_0 us}\mathrm{d}s\mathrm{d}x\mathrm{d}u. \tag{A.65}$$

Applying the derivative to the term in square brackets and using the representation of delta distribution again leads to

$$\langle u\rangle = \frac{\varepsilon_0 c_0}{4\pi P}$$
$$\times\iint \frac{1}{2\imath}\left[E^*\left(x+\frac{s}{2}\right)\partial_x E\left(x-\frac{s}{2}\right)-\partial_x E^*\left(x+\frac{s}{2}\right)E\left(x-\frac{s}{2}\right)\right]\delta\left(\frac{k_0 s}{2\pi}\right)\mathrm{d}s\mathrm{d}x, \tag{A.66}$$

which can be reduced to

$$\langle u\rangle = \frac{\varepsilon_0 c_0}{2k_0 P}\int \Im\{E^*(x)\partial_x E(x)\}\mathrm{d}x. \tag{A.67}$$

$\Im\{\cdot\}$ is the imaginary part.

- The second-order moment regarding x and u is defined by

$$\langle xu\rangle = \frac{\varepsilon_0 c_0 k_0}{4\pi P}\iiint E^*\left(x+\frac{s}{2}\right)E\left(x-\frac{s}{2}\right)e^{\imath k_0 us}xu\mathrm{d}s\mathrm{d}x\mathrm{d}u. \tag{A.68}$$

With the same calculations used for $\langle u\rangle$, the moment can be obtained directly:

$$\langle xu\rangle = \frac{\varepsilon_0 c_0}{2k_0 P}\int \Im\{E^*\partial_x E\}x\mathrm{d}x. \tag{A.69}$$

- The second-order moment regarding u is defined by

$$\langle u^2\rangle = \frac{\varepsilon_0 c_0 k_0}{4\pi P}\iiint E^*\left(x+\frac{s}{2}\right)E\left(x-\frac{s}{2}\right)e^{\imath k_0 us}u^2\mathrm{d}s\mathrm{d}x\mathrm{d}u. \tag{A.70}$$

With two-fold partial integration and the representation of the delta distribution the relation becomes

$$\langle u^2\rangle = \frac{\varepsilon_0 c_0}{4\pi k_0 P}$$
$$\times\iint \frac{1}{2}\partial_s\left[E^*\left(x+\frac{s}{2}\right)\partial_x E\left(x-\frac{s}{2}\right)-\partial_x E^*\left(x+\frac{s}{2}\right)E\left(x-\frac{s}{2}\right)\right]\delta\left(\frac{k_0 s}{2\pi}\right)\mathrm{d}s\mathrm{d}x. \tag{A.71}$$

Calculating the derivatives on E, the evaluating the integral regarding s gives

$$\langle u^2 \rangle = \frac{\varepsilon_0 c_0}{4\pi k_0 P} \times \int \frac{1}{4} \left[\partial_x E^*(x) \partial_x E(x) - E^*(x) \partial_x^2 E(x) + \partial_x E^*(x) \partial_x E(x) - \partial_x^2 E^*(x) E(x) \right] \mathrm{d}x. \tag{A.72}$$

Using partial integration again with $\int \partial_x^2 E^* E \mathrm{d}x = -\int \partial_x E^* \partial_x E \mathrm{d}x$ leads to

$$\langle u^2 \rangle = \frac{\varepsilon_0 c_0}{2k_0^2 P} \int |\partial_x E(x)|^2 \, \mathrm{d}x. \tag{A.73}$$

B. Material and Simulation Parameters

B.1. Parameters for Periodically Poled Nonlinear Crystals

Table B.1.: Material parameters for congruent lithium niobate with 5 mol % MgO doping.

symbol	description	value	unit
a_1	Sellmeier param. (extraordinary polarization) [148]	4.545 14	1
a_2	see above	0.096 471	μm^2
a_3	see above	0.043 763	μm^2
a_4	see above	0	μm^2
a_5	see above	0	μm^2
a_6	see above	0	μm^2
a_7	see above	0	μm^2
a_8	see above	−0.021 502	μm^{-2}
a_9	see above	0	μm^4
a_{10}	see above	0.4175	μm^3
a_{11}	see above	−0.6643	μm^2
a_{12}	see above	0.9036	μm
a_{13}	see above	3.5332	1
a_{14}	see above	−0.0744	μm^{-1}
a_{15}	see above	1×10^{-5}	$°C^{-1}$
a_{16}	see above	1.38×10^{-8}	$°C^{-2}$
T_a	see above	20	°C
α_{th}	thermal expansion factor (crystal axis X) [154]	15.4×10^{-4}	K^{-1}

Table B.2.: Material parameters for stoichiometric lithium tantalate with 1 mol % MgO doping.

symbol	description	value	unit
a_1	Sellmeier param. (extraordinary polarization) [149]	16.5658	1
a_2	see above	0.084 15	μm^2
a_3	see above	0.036 56	μm^2
a_4	see above	1.024×10^{-3}	μm^2
a_5	see above	0.0749	μm^2
a_6	see above	6428.69	μm^2
a_7	see above	532.75	μm^2
a_8	see above	0	μm^{-2}
a_9	see above	0.2367	μm^4
a_{10}	see above	−1.0951	μm^3
a_{11}	see above	2.2537	μm^2
a_{12}	see above	−1.2402	µm
a_{13}	see above	0.3395	1
a_{14}	see above	0	μm^{-1}
a_{15}	see above	1×10^{-5}	$°C^{-1}$
a_{16}	see above	1.71×10^{-8}	$°C^{-2}$
T_a	see above	22	°C
α_{th}	thermal expansion factor (crystal axis X) [154]	16.2×10^{-4}	K^{-1}

B.2. Parameters for Semiconductor Optical Amplifiers

Table B.3.: Layout and parameters of the vertical structure designed for an emission wavelength of 1120 nm.

description	thickness d/nm	refractive index n/1	thermal conductivity $\kappa/\mathrm{W\,m^{-1}\,K^{-1}}$
metal contact	5000	-	200
p-contact layer	100	3.459	44.0
p-cladding layer	60	3.360	15.9
p-cladding layer	1000	3.261	11.5
p-cladding layer	30	3.288	12.2
p-waveguide layer	800	3.316	13.2
	5	3.391	20.5
	2	3.459	44.0
quantum well	5	3.683	4.72
	2	3.459	44.0
	7.5	3.391	20.5
	2	3.459	44.0
quantum well	5	3.683	4.72
	2	3.459	44.0
	5	3.391	20.5
n-waveguide layer	4000	3.316	13.2
n-cladding layer	30	3.288	12.2
n-cladding layer	700	3.261	11.5
n-cladding layer	100	3.360	15.9
n-substrate	142 146	3.459	44.0

Table B.4.: Parameters used for the simulation of light propagating through ridge waveguide amplifiers.

symbol	description	value	unit
$\bar{n}$	reference refractive index	3.316	1
λ	wavelength	1120	nm
g'	differential modal gain	2039	m^{-1}
N_{tr}	transparency carrier density	1.82×10^{24}	m^{-3}
N_0	gain clamping carrier density	1.03×10^{24}	m^{-3}
α_0	internal background absorption	165	m^{-1}
f_N	modal cross section free carrier absorption	1.14×10^{-23}	m^2
r_s	series resistivity related to contact layer	3.802×10^{-9}	$\Omega\, m^2$
Ω_s	sheet resistance	17.95	$k\Omega$
Δn_0	built-in relative effective index		
	for $(x, z) \in$ etched trench	-4.8×10^{-3}	1
	else	0	1
n'_N	differential modal effective index	1.29×10^{-31}	m^3
n'_T	temperature induced differential index	2.5×10^{-4}	K^{-1}
E_g	band gab energy	1.13	eV
N_c	conduction band density of states	1.22×10^{24}	m^{-3}
N_v	valence band density of states	3.90×10^{24}	m^{-3}
n_0	equilibrium electron carrier density	1.00×10^{20}	m^{-3}
p_0	equilibrium hole carrier density	6.27×10^{2}	m^{-3}
μ_p	hole mobility in active region	3.67×10^{-2}	$m^2\, V^{-1}\, s^{-1}$
A_{SRH}	Shockly Read Hall recombination coefficient	9.79×10^{7}	s^{-1}
B_{SE}	spontaneous emission recomb. coefficient	1.37×10^{-16}	$m^3\, s^{-1}$
C_{Aug}	Auger recombination coefficient	4.00×10^{-42}	$m^6\, s^{-1}$
T_0	reference temperature	300	K

Table B.5.: Parameters used for the simulation of light propagating through tapered amplifiers.

symbol	description	value	unit
$\bar{n}$	reference refractive index	3.316	1
λ	wavelength	1120	nm
g'	differential modal gain	2039	m^{-1}
N_{tr}	transparency carrier density	1.82×10^{24}	m^{-3}
N_0	gain clamping carrier density	1.03×10^{24}	m^{-3}
α_0	internal background absorption	165	m^{-1}
f_N	modal cross section free carrier absorption	1.14×10^{-23}	m^2
r_s	series resistivity related to contact layer	3.802×10^{-9}	$\Omega\, m^2$
Ω_s	sheet resistance		
	for RW section	17.95	$k\Omega$
	for TP section	0.423	$k\Omega$
Δn_0	built-in relative effective index		
	for $(x, z) \in$ etched trench	-3.0×10^{-3}	1
	else	0	1
n'_N	differential modal effective index	1.29×10^{-31}	m^3
n'_T	temperature induced differential index	2.5×10^{-4}	K^{-1}
E_g	band gab energy	1.13	eV
N_c	conduction band density of states	1.22×10^{24}	m^{-3}
N_v	valence band density of states	3.90×10^{24}	m^{-3}
n_0	equilibrium electron carrier density	1.00×10^{20}	m^{-3}
p_0	equilibrium hole carrier density	6.27×10^{2}	m^{-3}
μ_p	hole mobility in active region	3.67×10^{-2}	$m^2\, V^{-1}\, s^{-1}$
A_{SRH}	Shockly Read Hall recombination coefficient	9.79×10^{7}	s^{-1}
B_{SE}	spontaneous emission recomb. coefficient	1.37×10^{-16}	$m^3\, s^{-1}$
C_{Aug}	Auger recombination coefficient	4.00×10^{-42}	$m^6\, s^{-1}$
T_0	reference temperature	300	K

Table B.6.: Elements of the ray transfer matrices of the optical setup for generation of three different feedback beam shapes.

feedback beam shape	$A/1$	B/m	C/m^{-1}	$D/1$
matched	2.907 174	$-5.095\,701 \times 10^{-3}$	$-1.462\,342 \times 10^{3}$	2.907 174
broad	−1.152 052	$4.603\,248 \times 10^{-3}$	$0.710\,852\,7 \times 10^{3}$	−1.152 052
narrow	1.712 463	$-1.482\,079 \times 10^{-3}$	$1.303\,932 \times 10^{3}$	1.712 463

B.3. Parameters for Distributed Bragg-Reflector Ridge Waveguide Lasers

Table B.7.: Parameters used for the simulation of distributed Bragg-reflector ridge waveguide lasers with external optical feedback. l_g: gain section length, R_f: front facet reflectivity.

symbol	description	value	unit
λ_l	central wavelength	1120	nm
n_g	group refractive index	3.6	1
α_H	linewidth enhancement factor	1.2	1
α_0	internal background absorption	200	m^{-1}
τ_{DBR}	DBR round-trip time	1×10^{-11}	s
R_{DBR}	DBR reflectance	0.35	1
τ_N	carrier lifetime	1.67×10^{-9}	s
I_{tr}	transparency current		
	for DBR-RWL with $l_g = 1\,mm$	13.0	mA
	for DBR-RWL with $l_g = 2\,mm$	26.0	mA
	for DBR-RWL with $l_g = 3\,mm$	39.0	mA
I_{th}	threshold current		
	for DBR-RWL with $l_g = 1\,mm$ and $R_f = 0.3$	46.8	mA
	for DBR-RWL with $l_g = 2\,mm$ and $R_f = 0.3$	55.1	mA
	for DBR-RWL with $l_g = 3\,mm$ and $R_f = 0.3$	57.8	mA
	for DBR-RWL with $l_g = 3\,mm$ and $R_f = 0.01$	98.0	mA
	for DBR-RWL with $l_g = 3\,mm$ and $R_f = 0.1$	65.1	mA

C. Measurement Methods and Devices

C.1. Laser Sources

C.1.1. External-Cavity Diode Laser

An external-cavity diode laser (ECDL) is used as a spectrally tunable laser source with emission wavelengths around 1110 nm. The laser source is realized by a ridge waveguide laser for external stabilization (RWLE) and an ECDL setup similar to [155]. A schematic of the RWLE is depicted in figure C.1a. The RWLE is based on an asymmetric super

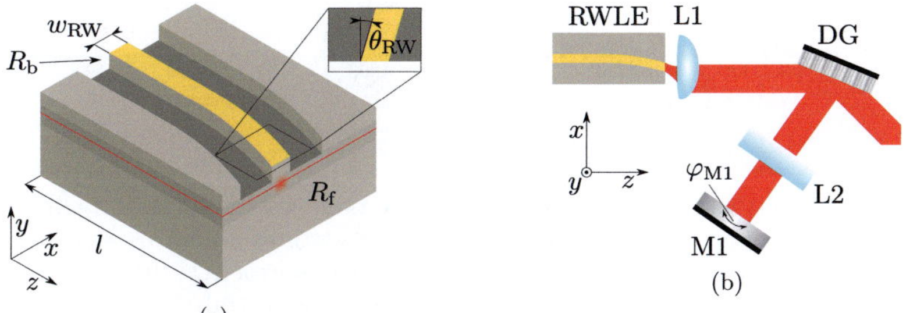

Figure C.1.: Schematic of (a) the ridge waveguide laser for external stabilization (RWLE) and (b) the setup of the external cavity. l: RWLE length, w_{RW}: ridge waveguide (RW) width, θ_{RW}: angle of RW relative to front facet, R_b: back facet reflectivity, R_f: front facet reflectivity, L1: aspherical lens, L2: cylindrical lens, DG: diffraction grating, M1: plane mirror, φ_{M1}: angle of M1.

large optical cavity, which is described in detail in section 4.2. The RWLE has a length of $l = 4$ mm and the horizontal waveguiding is achieved by a ridge waveguide (RW) with a width of $w_{RW} = 4$ µm. This results in a refractive index step of -3×10^{-3}. While the back facet is highly reflective with a reflectivity of $R_b = 0.96$, the front facet is antireflective coated with a residual reflectivity of $R_f = 3 \times 10^{-4}$. In order to further decrease the effective facet reflectivity, the RW has an outcoupling bend to realize an angle at the front facet of $\theta_{RW} = 3°$.

The setup of the external cavity for the RWLE is shown in figure C.1b. An aspherical lens L1 with a focal length of 8 mm collimates the emission from the RWLE. The beam is diffracted in first-order at the diffraction grating (DG) with a grating constant of $1200\,\text{mm}^{-1}$ and passes through a cylindrical lens L2 with a focal length of 60 mm. The lens L2 focuses the beam in the vertical (y) direction onto a plane mirror M1. The angle of the mirror φ_{M1} can be used to adjust the emission wavelength of the RWLE. Changing the angle by 1° shifts the emission spectrum by 12.5 nm. The laser emission for the experiments is coupled out via the zero-order diffraction at the DG.

During the experiments, the RWLE is operated at a temperature of 25 °C and a constant injection current of 400 mA. The exemplary emission characteristics of the ECDL are shown in figure C.2. At all used working points in the range between 1090 nm

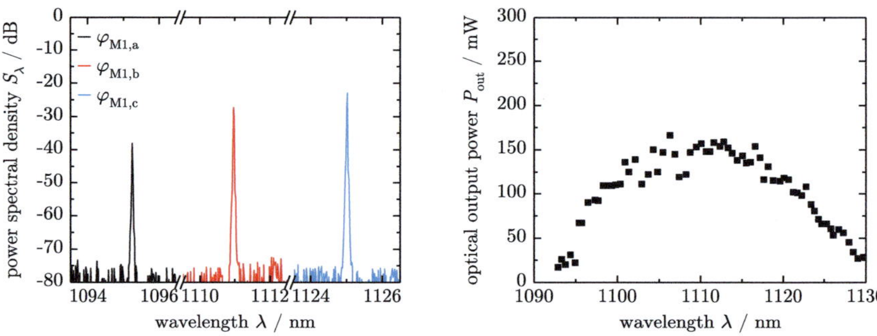

(a) (b)

Figure C.2.: Output emission characteristics of the ridge waveguide laser stabilized by an external cavity. (a) Emission spectra at three exemplary mirror angles $\varphi_{M1,a}$, $\varphi_{M1,b}$ and $\varphi_{M1,c}$ of the external cavity mirror M1. (b) Optical output power of the ECDL system as a function of the central emission wavelength.

and 1130 nm the emission spectra are characterized by a single spectral lobe, while the output powers are in the range between 20 mW and 150 mW.

C.1.2. Distributed Bragg-Reflector Ridge Waveguide Laser

Distributed Bragg-reflector ridge waveguide lasers (DBR-RWL) are used as laser light sources with emission wavelengths around 1064 nm, 1120 nm, and 1178 nm. Figure C.3a shows the schematic of the DBR-RWL. The DBR-RWLs have a total length of 4 mm, which is divided into the distributed Bragg-reflector (DBR) section with a length of $l_{DBR} = 1$ mm and a gain section with a length of $l_g = 3$ mm. The ridge waveguide (RW) with a width of $w_{RW} = 4$ µm is formed by two etched trenches that result in an effective index step of -3×10^{-3}. While the back facet is antireflective, the front facet is coated to achieve a specific reflectivity R_f. The actual reflectivity is chosen with respect to the individual DBR-RWL characteristics and the requirements of the experiment. The front facet reflectivities of the three DBR-RWLs with design wavelengths of 1064 nm, 1120 nm, and 1178 nm are 0.01 (1 %), 0.3 (30 %) and 0.1 (10 %), respectively. The DBR-RWL with an emission wavelength of 1178 nm also has metal heating wires and

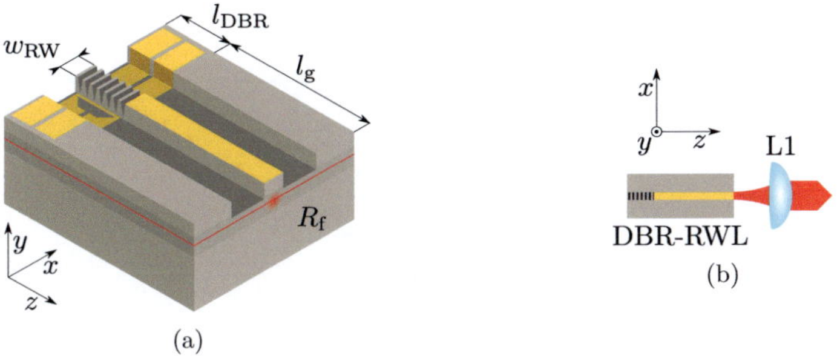

Figure C.3.: Schematic of (a) the distributed Bragg-reflector ridge waveguide laser (DBR-RWL) and (b) the setup used for the experiments. l_{g}: gain section length, l_{DBR}: distributed Bragg-reflector section length, w_{RW}: ridge waveguide width, R_{f}: front facet reflectivity, L1: aspherical lens.

corresponding contact pads in the DBR section.

The emission from the DBR-RWLs is collimated by an aspherical lens, which is shown schematically in figure C.3b. For the DBR-RWL with design wavelengths of 1064 nm and 1120 nm, an aspherical lens with a focal length of 8 mm is used, while a lens with a focal length of 4.5 mm is chosen for the DBR-RWL with a design wavelength of 1178 nm.

In figure C.4 the output emission characteristics of the DBR-RWL with a design wavelength of 1064 nm is shown. While the DBR-RWL is operated at a constant injection current of 250 mA, the temperature is varied between 10 °C and 60 °C. Thereby, the central emission wavelength can be tuned continuously in a range between 1062.6 nm and 1066.3 nm without facing spectral mode hops. The emission spectra at three temperatures T are shown in figure C.4a where only a single spectral mode is observed. In figure C.4b the emitted output power is plotted at exemplary working points within the tuning range. The output power remains in a range between 150 mW and 300 mW over the entire spectral tuning range.

The DBR-RWL with a design wavelength of 1178 nm is also used as a laser source with a tunable emission spectrum. Here, the heater on the DBR section is utilized. By applying a heating current, the local temperature of the DBR can be increased, resulting in a change of the emission wavelength. This allows the central emission wavelength to be tuned by up to 3 nm. In order to further change the central emission wavelength, the DBR-RWL is operated at different temperatures of either 15 °C or 25 °C. A tuning range of the emission spectrum from 1174.0 nm to 1177.5 nm can be realized. Three exemplary emission spectra at working points with a constant default injection of 200 mA but different temperatures T and heating currents I_{DBR} are shown in figure C.5a. Again, only a single spectral mode is observed. The corresponding output power of the DBR-RWL at central emission wavelengths between 1174.0 nm and 1177.5 nm is given in figure C.5b. Neither the temperature nor the heating current has a significant effect on the output power, which remains nearly constant around 60 mW.

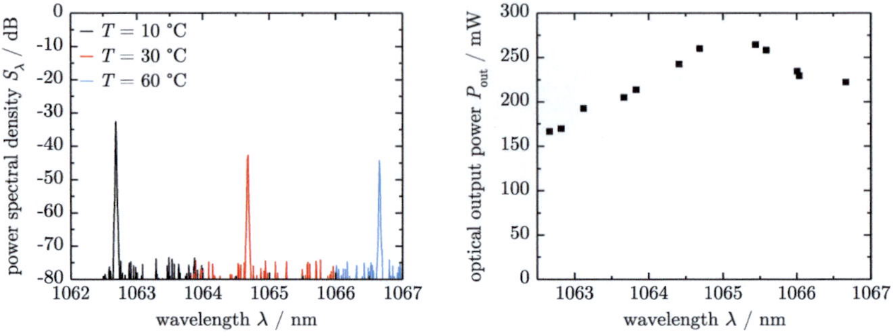

(a) (b)

Figure C.4.: Output emission characteristics of the distributed Bragg-reflector ridge waveguide laser with a design wavelength of 1064 nm at a constant injection current of 250 mA. (a) Exemplary emission spectra at three different temperatures T. (b) Optical output power as a function of the central emission wavelength.

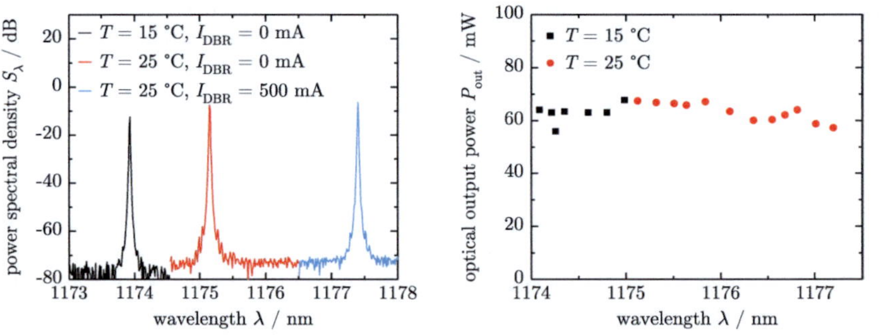

(a) (b)

Figure C.5.: Output emission characteristics of the distributed Bragg-reflector ridge waveguide laser with a design wavelength of 1178 nm at a constant injection current of 200 mA. (a) Exemplary emission spectra at three different working points with varying temperature T and heater current I_{DBR}. (b) Optical output power as a function of the central emission wavelength.

In contrast to the previously presented laser sources, the DBR-RWL with a design wavelength of 1120 nm is used as a laser source with adjustable output power. The output power is directly tuned by the injection current I and the corresponding power-current characteristic is given in figure C.6a. The DBR-RWL is operated at a default

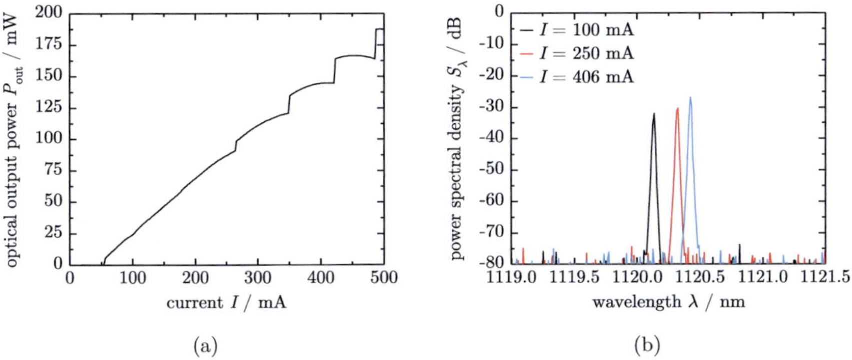

Figure C.6.: Output emission characteristics of the distributed Bragg-reflector ridge waveguide laser with a design wavelength of 1120 nm. (a) Optical output power as a function of injection current. (b) Exemplary emission spectra at three different working points with varying injection current I.

temperature of 25 °C. While the laser threshold is reached at a current around 50 mA, more than 185 mW output power is achieved at the maximum current of 500 mA. Several kinks due to spectral mode hops can be distinguished in the power-current characteristics. In figure C.6b the emission spectra at three different injection currents are presented. Only a single spectral mode is observed at each working point where the spectral shift due to self-heating and mode hops is below 0.5 nm.

C.2. Optical Power Characterization

The optical power is measured with detectors from *Gentec EO*. Depending on the actual power level and wavelength of the light to be measured, different power detectors are used. The different detectors and their specifications are listed in table C.1. The detectors are used in combination with the monitoring devices *Gentec EO Solo 2* and *Uno*. The total uncertainty of the power measurement is thereby increased by ±0.5 % of the selected measuring range of the monitoring device.

Table C.1.: List of the power detectors used from *Gentec EO* and their specifications.

detector model	detection principle	spectral range	maximum power	relative uncertainty
PH100-Si	silicon	$320\,\text{nm} \leq \lambda \leq 1100\,\text{nm}$	36 mW	±2.5 %
PH20-Ge	germanium	$800\,\text{nm} \leq \lambda \leq 1650\,\text{nm}$	30 mW	±3.5 %
PH20-Ge-OD2	germanium	$800\,\text{nm} \leq \lambda \leq 1650\,\text{nm}$	500 mW	±5 %
XLP12-3S-H2	thermopile	$280\,\text{nm} \leq \lambda \leq 2100\,\text{nm}$	3 W	±2.5 %
UP19K-15S-W5	thermopile	$190\,\text{nm} \leq \lambda \leq 10\,\mu\text{m}$	15 W	±2.5 %

C.3. Emission Spectrum Characterization

The optical spectrum is recorded with optical spectrum analyzers (OSA) depending on the wavelength range investigated. For wavelengths below 1090 nm a system from *LTB Lasertechnik Berlin GmbH* based on the DEMON monochromator is used. The spectral resolution at 1000 nm is 12 pm, while the absolute accuracy at this wavelength is 4 pm. A dynamic range of 16 bit is specified for the detector.

For experiments with emission wavelengths above 1090 nm the *Q8384* OSA from *Advantest Corporation* is used. The OSA achieves a resolution of 10 pm with an accuracy of 20 pm and a dynamic range of 60 dB.

To determine the center wavelength and width of an emission spectrum containing a single lobe, the measured spectrum is fitted with a Gaussian. In contrast, for more complex emission spectra containing multiple lobes, the center wavelength and spectral width are calculated as the median wavelength and full width at half maximum, respectively. Throughout this work, the measured wavelengths are always given as vacuum wavelengths.

Precise measurements of the spectral shift and spectral width that are below the resolution limit of the OSAs are performed with a Fabry-Pérot interferometer *FPI 100-0980-3V0* from *TOPTICA Photonics AG*. With a free spectral range of 1 GHz the device has a resolution of 2.0 MHz.

C.4. Spatial Beam Characterization

The spatial emission properties are characterized by measuring the near field and far field intensity distributions. Throughout this work, the term near field is used to refer to the spatial intensity distribution within or at the facet of, for example, a crystal or laser diode. The setup for the measurements is shown in figure C.7. The light first passes through the lens L1, which is usually part of the respective experimental setup and has different geometries and focal lengths f_{L1}. The spherical lens L2 with a focal length of $f_{\text{L2}} = 1000\,\text{mm}$ is optional and only needed for the measurement of the far field intensity distribution. The spherical lens L3 with a focal length of $f_{\text{L3}} = 400\,\text{mm}$ is used to image the respective intensity distribution onto a camera (CAM). The camera is placed in the focal plane behind L3.

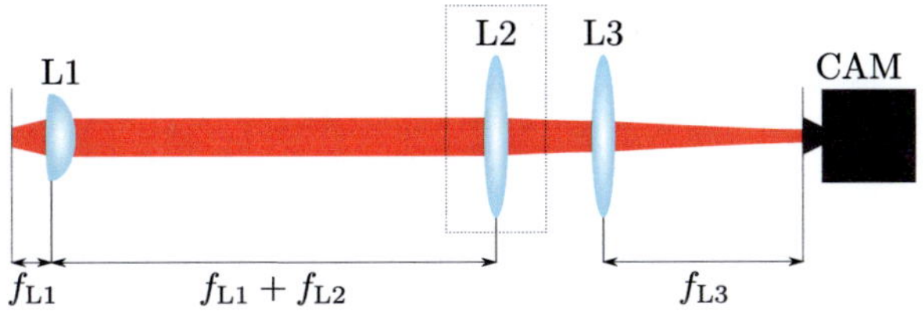

Figure C.7.: Schematic of the setup for the measurement of the near field and far field intensity distributions. L1: lens (actual geometry depends on setup), L2: optional spherical lens, L3: spherical lens, CAM: camera, f_{L1} - f_{L3}: focal length of L1 to L3.

To record the near field intensity distribution, L2 is removed. Thus, the lenses L1 and L3 form a stretched telescope and the intensity distribution in the focal plane in front of L1 is imaged. The corresponding magnification $M_{L1,L3}$ is given by

$$M_{L1,L3} = f_{L3}/f_{L1}. \tag{C.1}$$

Depending on the actual optical setup, the optical path from the beam source to the focal plane of the lens L1 can contain additional lenses. Thus, the imaged near field intensity distribution at the focal plane of the lens L1 can already be an image. Assuming that an addition lens with a focal length of f_{L0} is used to image the beam waist of the source to the focal plane of the lens L1, the additional magnification must be considered for the total magnification. For imaging fulfilling the lens equation, the magnification M_{L0} is defined by

$$M_{L0} = \frac{f_{L0}}{\Delta a}, \tag{C.2}$$

where Δa is the distance of the source to the focal plane of lens L0. The total magnification is then given as the product of the individual magnifications.

For the record of the far field intensity distribution, L2 is placed between L1 and L3. The distance to lens L1 is chosen as the sum of their focal lengths. Consequently, the far field angles $\theta_{x,y}$ are related to the spatial coordinates of the camera $d_{x,y}$ by

$$\theta_{x,y} = \frac{f_{L2}}{f_{L1}} \arctan\left(\frac{d_{x,y}}{f_{L3}}\right). \tag{C.3}$$

Two different cameras manufactured by *Point Grey Research* with either CCD or CMOS sensors are used for the measurements. The CCD camera *Scorpion SCOR-20-SOM* has a sensor resolution of 1600×1200 pixels with a pixel size of 4.4 µm. A dynamic range of 12 bit is specified. The CMOS camera *Grasshopper3 GS3-U3-91S6M* features a sensor resolution of 3376×2704 pixels and a pixel size of 3.69 µm. Here, the dynamic range is specified with 14 bit.

To measure the intensity distributions, a dynamic range of 8 bit is used, while eight images are averaged to increase the signal-to-noise ratio. The intensity distributions are normalized to either the total measured optical power or the maximum intensity. Throughout this work the width of an intensity distribution in the near field or far field is defined as the $1/e^2$ threshold width with respect to the maximum intensity.

Bibliography

[1] M. Haverkamp, G. Kochem, and K. Boucke, „Single mode fiber coupled tapered laser module with frequency stabilized spectrum“, in *High-Power Diode Laser Technology and Applications VI*, vol. 6876, SPIE, 2008, p. 68761D.

[2] D. Jedrzejczyk, A. Sahm, C. Carstens, *et al.*, „Coupling of a high-power tapered diode laser beam into a single-mode-fiber within a compact module“, in *High-Power Diode Laser Technology and Applications XIII*, vol. 9348, SPIE, 2015, p. 93480Z.

[3] J. Wang, X. Zhu, R. A. Norwood, and N. Peyghambarian, „Beyond 3 µm Dy^{3+} / Er^{3+} co-doped ZBLAN fiber lasers pumped by 976 nm laser diode“, *Applied Physics Letters*, vol. 118, no. 15, p. 151 101, 2021.

[4] A. C. Wilson, J. C. Sharpe, C. R. McKenzie, P. J. Manson, and D. M. Warrington, „Narrow-linewidth master-oscillator power amplifier based on a semiconductor tapered amplifier“, *Applied Optics*, vol. 37, no. 21, pp. 4871–4875, 1998.

[5] S. Spießberger, M. Schiemangk, A. Sahm, *et al.*, „Micro-integrated 1 watt semiconductor laser system with a linewidth of 3.6 kHz“, *Optics Express*, vol. 19, no. 8, pp. 7077–7083, 2011.

[6] D. Brauda, H. Thiem, M. Schütz, and M. Hager, „Low SWaP miniaturized fiber coupled external cavity diode laser (ECDL) with high optical isolation as plug and play solution“, in *Components and Packaging for Laser Systems IX*, vol. 12402, SPIE, 2023, p. 1 240 204.

[7] Z. Fu, P. Wang, S. Chai, L. Huang, and J. Zhang, „Bose-Einstein condensate in a light-induced vector gauge potential using 1064-nm optical-dipole-trap lasers“, *Physical Review A*, vol. 84, p. 043 609, 2011.

[8] K. A. Fedorova, G. S. Sokolovskii, M. Khomylev, D. A. Livshits, and E. U. Rafailov, „Efficient yellow-green light generation at 561 nm by frequency-doubling of a QD-FBG laser diode in a PPLN waveguide“, *Optics Letters*, vol. 39, no. 23, pp. 6672–6674, 2014.

[9] J. Hofmann, A. Sahm, W. John, F. Bugge, and K. Paschke, „Generation of second harmonic light with a wavelength of 560 nm in a compact module“, *Optics & Laser Technology*, vol. 83, pp. 55–58, 2016.

[10] „532, 561, 594 nm compact visible lasers“, QD Laser, Inc., white paper. [Online]. Available: `https://www.qdlaser.com/assets/pdf/product/DP0067-01_White%20paper%20Compact%20visible%20lasers.pdf` (visited on 03/02/2023).

[11] K. Yamagata, R. Suetsugu, and T. Wakayama, „Long-term, six-dimensional live-cell imaging for the mouse preimplantation embryo that does not affect full-term development“, *Journal of Reproduction and Development*, vol. 55, no. 3, pp. 343–350, 2009.

[12] K. Elgass, D. Hartmann, J. Siebenmorgen, T. Kalkbrenner, and K. Weisshart, „Breakthrough instruments and products: Lattice Lightsheet 7 - non-invasive imaging of the subcellular dynamics of life“, *Review of Scientific Instruments*, vol. 92, no. 10, p. 109 501, 2021.

[13] R. Broom, E. Mohn, C. Risch, and R. Salathe, „Microwave self-modulation of a diode laser coupled to an external cavity“, *IEEE Journal of Quantum Electronics*, vol. 6, no. 6, pp. 328–334, 1970.

[14] S. Schwertfeger, A. Klehr, G. Erbert, and G. Tränkle, „Compact hybrid master oscillator power amplifier with 3.1-W CW output power at wavelengths around 1061 nm“, *IEEE Photonics Technology Letters*, vol. 16, no. 5, pp. 1268–1270, 2004.

[15] R. Bege, D. Jedrzejczyk, G. Blume, *et al.*, „Reduction of optical feedback originating from ferroelectric domains of periodically poled crystals“, *IEEE Journal of Quantum Electronics*, vol. 53, no. 5, pp. 1–9, 2017.

[16] J. H. Marquardt, F. C. Cruz, M. Stephens, *et al.*, „Grating-tuned semiconductor MOPA lasers for precision spectroscopy“, in *Application of Tunable Diode and Other Infrared Sources for Atmospheric Studies and Industrial Process Monitoring*, vol. 2834, SPIE, 1996, pp. 34–40.

[17] T. Tekin, H. Schroder, B. Wunderle, *et al.*, „A compact integrated green-light source by second harmonic generation of a GaAs distributed feedback laser diode“, in *Micro-Optics 2008*, vol. 6992, SPIE, 2008, p. 69920O.

[18] A. K. Hansen, M. Christensen, D. Noordegraaf, *et al.*, „Efficient generation of 1.9 W yellow light by cascaded frequency doubling of a distributed Bragg reflector tapered diode“, *Applied Optics*, vol. 55, no. 32, pp. 9270–9274, 2016.

[19] J. Hofmann, G. Blume, D. Jedrzejczyk, *et al.*, „Miniaturized diode laser module emitting green light at 532 nm with a power of more than 900 mW for next-generation holographic displays“, *Optical Review*, vol. 23, no. 1, pp. 141–145, 2016.

[20] E. Kapon, *Semiconductor Lasers II*, 1st ed. Academic Press, 1999.

[21] K. Paschke, J. Pohl, D. Feise, G. Blume, and G. Erbert, „Properties of 62x nm red-emitting single mode diode lasers“, in *Novel In-Plane Semiconductor Lasers XIII*, vol. 9002, SPIE, 2014, p. 90020A.

[22] F. Bugge, M. Zorn, U. Zeimer, *et al.*, „Highly strained very high-power laser diodes with InGaAs QWs“, *Journal of Crystal Growth*, vol. 248, pp. 354–358, 2003.

[23] P. Ressel, G. Erbert, U. Zeimer, *et al.*, „Novel passivation process for the mirror facets of Al-free active-region high-power semiconductor diode lasers“, *IEEE Photonics Technology Letters*, vol. 17, no. 5, pp. 962–964, 2005.

[24] J. E. Boschker, U. Spengler, P. Ressel, M. Schmidbauer, A. Mogilatenko, and A. Knigge, „Stability of ZnSe-passivated laser facets cleaved in air and in ultra-high vacuum“, *IEEE Photonics Journal*, vol. 14, no. 3, pp. 1–6, 2022.

[25] G. Blume, M. Schiemangk, J. Pohl, *et al.*, „Narrow linewidth of 633-nm DBR ridge-waveguide lasers“, *IEEE Photonics Technology Letters*, vol. 25, no. 6, pp. 550–552, 2013.

[26] P. Friedmann, J. Schleife, J. Gilly, and M. T. Kelemen, „High efficiency laser sources usable for single mode fiber coupling and frequency doubling“, in *Novel In-Plane Semiconductor Lasers XI*, vol. 8277, SPIE, 2012, p. 82771M.

[27] A. Sahm, N. Werner, J. Hofmann, D. Jedrzejczyk, D. Feise, and K. Paschke, „Miniaturized watt-level laser modules emitting in the yellow-green spectral range for biophotonic applications“, in *Nonlinear Frequency Generation and Conversion: Materials and Devices XVIII*, vol. 10902, SPIE, 2019, p. 109020C.

[28] R. W. Boyd, *Nonlinear Optics*, 4th ed. Academic Press, 2020.

[29] D. S. Hum and M. M. Fejer, „Quasi-phasematching“, *Comptes Rendus Physique*, vol. 8, no. 2, pp. 180–198, 2007.

[30] M. M. Fejer, G. A. Magel, D. H. Jundt, and R. L. Byer, „Quasi-phase-matched second harmonic generation: Tuning and tolerances“, *IEEE Journal of Quantum Electronics*, vol. 28, no. 11, pp. 2631–2654, 1992.

[31] D. Fluck, „Theory on phase-matched second-harmonic generation in biaxial planar waveguides“, *IEEE Journal of Quantum Electronics*, vol. 35, no. 1, pp. 53–59, 1999.

[32] D. Jedrzejczyk, R. Güther, K. Paschke, G. Erbert, and G. Tränkle, „Diode laser frequency doubling in a ppMgO:LN ridge waveguide: Influence of structural imperfection, optical absorption and heat generation“, *Applied Physics B*, vol. 109, no. 1, pp. 33–42, 2012.

[33] G. D. Boyd and D. A. Kleinman, „Parametric interaction of focused Gaussian light beams“, *Journal of Applied Physics*, vol. 39, no. 8, pp. 3597–3639, 1968.

[34] J. A. Armstrong, N. Bloembergen, J. Ducuing, and P. S. Pershan, „Interactions between light waves in a nonlinear dielectric“, *Physical Review*, vol. 127, no. 6, pp. 1918–1939, 1962.

[35] R. C. Miller, „Optical harmonic generation in single crystal $BaTiO_3$“, *Physical Review*, vol. 134, no. 5A, pp. A1313–A1319, 1964.

[36] E. J. Lim, M. M. Fejer, and R. L. Byer, „Second-harmonic generation of green light in periodically poled planar lithium niobate waveguide“, *Electronics Letters*, vol. 25, no. 3, pp. 174–175, 1989.

[37] C. J. van der Poel, J. D. Bierlein, J. B. Brown, and S. Colak, „Efficient type I blue second-harmonic generation in periodically segmented $KTiOPO_4$ waveguides“, *Applied Physics Letters*, vol. 57, no. 20, pp. 2074–2076, 1990.

[38] S. Matsumoto, E. J. Lim, H. M. Hertz, and M. M. Fejer, „Quasiphase-matched second harmonic generation of blue light in electrically periodically-poled lithium tantalate waveguides“, *Electronics Letters*, vol. 27, no. 22, pp. 2040–2042, 1991.

[39] M. Yamada, N. Nada, M. Saitoh, and K. Watanabe, „First-order quasi-phase matched $LiNbO_3$ waveguide periodically poled by applying an external field for efficient blue second-harmonic generation“, *Applied Physics Letters*, vol. 62, no. 5, pp. 435–436, 1993.

[40] K. Shinozaki, T. Fukunaga, K. Watanabe, and T. Kamijoh, „Self-quasi-phase-matched second-harmonic generation in the proton-exchanged $LiNbO_3$ optical waveguide with periodically domain-inverted regions“, *Applied Physics Letters*, vol. 59, no. 5, pp. 510–512, 1991.

[41] W. P. Risk and S. D. Lau, „Distributed-Bragg-reflection properties of segmented KTP waveguides“, *Optics Letters*, vol. 18, no. 4, p. 272, 1993.

[42] M. Oron, M. Katz, A. Zussman, D. Eger, and A. Shahna, „Bragg reflection in periodically segmented $KTiOPO_4$ waveguides“, *Journal of Applied Physics*, vol. 76, no. 11, pp. 7519–7523, 1994.

[43] T. J. Yang, V. Gopalan, P. Swart, and U. Mohideen, „Experimental study of internal fields and movement of single ferroelectric domain walls“, *Journal of Physics and Chemistry of Solids*, vol. 61, no. 2, pp. 275–282, 2000.

[44] S. Kim and V. Gopalan, „Optical index profile at an antiparallel ferroelectric domain wall in lithium niobate“, *Materials Science and Engineering: B*, vol. 120, no. 1, pp. 91–94, 2005.

[45] Y. Chen, H. Zhan, and B. Zhou, „Refractive index modulation in periodically poled MgO-doped congruent $LiNbO_3$ crystal“, *Applied Physics Letters*, vol. 93, no. 22, p. 222 902, 2008.

[46] S.-C. Pei, T.-S. Ho, C.-C. Tsai, *et al.*, „Non-invasive characterization of the domain boundary and structure properties of periodically poled ferroelectrics“, *Optics Express*, vol. 19, no. 8, p. 7153, 2011.

[47] J. Golde, M. Rüsing, J. Rix, L. M. Eng, and E. Koch, „Quantifying the refractive index of ferroelectric domain walls in periodically poled $LiNbO_3$ single crystals by polarization-sensitive optical coherence tomography“, *Optics Express*, vol. 29, no. 21, p. 33 615, 2021.

[48] T. Jach, S. Kim, V. Gopalan, S. Durbin, and D. Bright, „Long-range strains and the effects of applied field at 180° ferroelectric domain walls in lithium niobate“, *Physical Review B*, vol. 69, no. 6, p. 064 113, 2004.

[49] V. Gopalan, V. Dierolf, and D. A. Scrymgeour, „Defect–domain wall interactions in trigonal ferroelectrics“, *Annual Review of Materials Research*, vol. 37, no. 1, pp. 449–489, 2007.

[50] E. Autizi, A.-D. Capobianco, F. M. Pigozzo, *et al.*, „Modeling and characterization of border domain effects in periodically poled lithium niobate crystals grown by the off-center Czochralski technique“, *Optical and Quantum Electronics*, vol. 38, no. 1, pp. 177–185, 2006.

[51] J. W. Choi, H. Kang, D.-K. Ko, J. H. Ro, and N. E. Yu, „Device quality evaluation of periodically poled crystals by diffraction analysis at different wavelengths“, *Japanese Journal of Applied Physics*, vol. 52, no. 2R, p. 026 701, 2013.

[52] C. Moser, F. Havermeyer, W. Liu, G. Steckman, K. Buse, and D. Psaltis, „Volume Bragg grating devices“, in *OFC 2003 Optical Fiber Communications Conference, 2003.*, Optica Publishing Group, 2003, pp. 644–645.

[53] J. E. Hellstrom, B. Jacobsson, V. Pasiskevicius, and F. Laurell, „Finite beams in reflective volume Bragg gratings: Theory and experiments“, *IEEE Journal of Quantum Electronics*, vol. 44, no. 1, pp. 81–89, 2008.

[54] K. O. Hill, Y. Fujii, D. C. Johnson, and B. S. Kawasaki, „Photosensitivity in optical fiber waveguides: Application to reflection filter fabrication“, *Applied Physics Letters*, vol. 32, no. 10, pp. 647–649, 1978.

[55] T. Erdogan, „Fiber grating spectra“, *Journal of Lightwave Technology*, vol. 15, no. 8, pp. 1277–1294, 1997.

[56] S. Kurimura, Y. Kato, M. Maruyama, Y. Usui, and H. Nakajima, „Quasi-phase-matched adhered ridge waveguide in $LiNbO_3$“, *Applied Physics Letters*, vol. 89, no. 19, p. 191 123, 2006.

[57] Y. Gao, J. Sun, G. Chen, and C. Sima, „Demonstration of simultaneous mode conversion and demultiplexing for mode and wavelength division multiplexing systems based on tilted few-mode fiber Bragg gratings“, *Optics Express*, vol. 23, no. 8, pp. 9959–9967, 2015.

[58] P. Guan, M. Tang, M. Cao, *et al.*, „Transverse asymmetry of the index modulation profile in few-mode fiber Bragg grating“, *Photonics*, vol. 8, no. 3, p. 87, 2021.

[59] E. Bradley, P. Yu, and R. Mathis, „Control of Bragg grating resonant wavelength through its dependence on the effective index of a waveguide“, *Journal of Lightwave Technology*, vol. 15, no. 7, pp. 1156–1164, 1997.

[60] H. Kogelnik, „Coupled wave theory for thick hologram gratings“, *The Bell System Technical Journal*, vol. 48, no. 9, pp. 2909–2947, 1969.

[61] H. Wenzel, R. Güther, A. M. Shams-Zadeh-Amiri, and P. Bienstman, „A comparative study of higher order Bragg gratings: Coupled-mode theory versus mode expansion modeling“, *IEEE Journal of Quantum Electronics*, vol. 42, no. 1, pp. 64–70, 2006.

[62] L. A. Coldren, S. W. Corzine, and M. L. Masanovic, *Diode Lasers and Photonic Integrated Circuits*, 2nd ed. John Wiley & Sons, Ltd, 2012.

[63] J. B. Shellan, P. Agmon, P. Yeh, and A. Yariv, „Statistical analysis of Bragg reflectors*", *Journal of the Optical Society of America*, vol. 68, no. 1, pp. 18–27, 1978.

[64] H.-T. Hsieh, W. Liu, F. Havermeyer, C. Moser, and D. Psaltis, „Beam-width-dependent filtering properties of strong volume holographic gratings", *Applied Optics*, vol. 45, no. 16, p. 3774, 2006.

[65] J. Albert, S. Theriault, F. Bilodeau, *et al.*, „Minimization of phase errors in long fiber Bragg grating phase masks made using electron beam lithography", *IEEE Photonics Technology Letters*, vol. 8, no. 10, pp. 1334–1336, 1996.

[66] G. Coppola, A. Irace, A. Cutolo, and M. Iodice, „Effect of fabrication errors in channel waveguide Bragg gratings", *Applied Optics*, vol. 38, no. 9, p. 1752, 1999.

[67] R. Feced and M. Zervas, „Effects of random phase and amplitude errors in optical fiber Bragg gratings", *Journal of Lightwave Technology*, vol. 18, no. 1, pp. 90–101, 2000.

[68] J. Hofmann, N. Werner, D. Feise, *et al.*, „Comparison of yellow light emitting micro integrated laser modules with different geometries of the crystals for second harmonic generation", in *Nonlinear Frequency Generation and Conversion: Materials, Devices, and Applications XV*, vol. 9731, SPIE, 2016, p. 973 109.

[69] R. Bege, D. Jedrzejczyk, J. Hofmann, *et al.*, „Watt-level yellow emitting lasers by frequency doubling of high power diode lasers", in *High-Power, High-Energy, and High-Intensity Laser Technology III*, vol. 10238, SPIE, 2017, p. 102380X.

[70] S. Watanabe, J. Hirohashi, K. Imai, M. Hoshi, and S. Makio, „Femtosecond-laser-written circular waveguides in MgO-doped stoichiometric $LiTaO_3$", in *Conference on Lasers and Electro-Optics (CLEO)*, Optica Publishing Group, 2019, p. JTu2A.106.

[71] S. Watanabe, J. Hirohashi, K. Imai, M. Hoshi, and S. Makio, „Femtosecond-laser-written waveguides in periodically poled MgO-doped stoichiometric lithium tantalate for frequency conversion", in *2019 Conference on Lasers and Electro-Optics Europe and European Quantum Electronics Conference (CLEO/Europe-EQEC)*, IEEE, 2019, p. 1.

[72] M. Connelly, „Wideband semiconductor optical amplifier steady-state numerical model", *IEEE Journal of Quantum Electronics*, vol. 37, no. 3, pp. 439–447, 2001.

[73] F. Rinner, J. Rogg, P. Friedmann, M. Mikulla, G. Weimann, and R. Poprawe, „Longitudinal carrier density measurement of high power broad area laser diodes", *Applied Physics Letters*, vol. 80, no. 1, pp. 19–21, 2002.

[74] Z. Dai, R. Michalzik, P. Unger, and K. Ebeling, „Numerical simulation of broad-area high-power semiconductor laser amplifiers", *IEEE Journal of Quantum Electronics*, vol. 33, no. 12, pp. 2240–2254, 1997.

[75] E. Zhou, X. Zhang, and D. Huang, „Analysis on dynamic characteristics of semiconductor optical amplifiers with certain facet reflection based on detailed wideband model“, *Optics Express*, vol. 15, no. 14, p. 9096, 2007.

[76] Z. Dai, R. Michalzik, P. Unger, and K. Ebeling, „Impact of the residual facet reflectivity on beam profile filamentation in semiconductor laser amplifiers“, in *Summaries of papers presented at the Conference on Lasers and Electro-Optics*, IEEE, 1996, pp. 416–417.

[77] M. Mikulla, P. Chazan, A. Schmitt, *et al.*, „High-brightness tapered semiconductor laser oscillators and amplifiers with low-modal gain epilayer-structures“, *IEEE Photonics Technology Letters*, vol. 10, no. 5, pp. 654–656, 1998.

[78] S. Sujecki, L. Borruel, J. Wykes, *et al.*, „Nonlinear properties of tapered laser cavities“, *IEEE Journal of Selected Topics in Quantum Electronics*, vol. 9, no. 3, pp. 823–834, 2003.

[79] M. A. Helal, S. N. Nyirenda-Kaunga, S. Bull, and E. C. Larkins, „Beam quality degradation processes in tapered lasers and DBR tapered lasers“, in *2017 IEEE High Power Diode Lasers and Systems Conference (HPD)*, IEEE, 2017, pp. 25–26.

[80] T. Liu, K. Obermann, K. Petermann, F. Girardin, and G. Guekos, „Effect of saturation caused by amplified spontaneous emission on semiconductor optical amplifier performance“, *Electronics Letters*, vol. 33, no. 24, pp. 2042–2043, 1997.

[81] G. Talli and M. J. Adams, „Amplified spontaneous emission in semiconductor optical amplifiers: Modelling and experiments“, *Optics Communications*, vol. 218, no. 1, pp. 161–166, 2003.

[82] A. Collar, G. Henshall, J. Farre, *et al.*, „Low residual reflectivity of angled-facet semiconductor laser amplifiers“, *IEEE Photonics Technology Letters*, vol. 2, no. 8, pp. 553–555, 1990.

[83] P.-A. Besse, J.-S. Gu, and H. Melchior, „Reflectivity minimization of semiconductor laser amplifiers with coated and angled facets considering two-dimensional beam profiles“, *IEEE Journal of Quantum Electronics*, vol. 27, no. 6, pp. 1830–1836, 1991.

[84] A. Klehr, H. Wenzel, J. Fricke, F. Bugge, and G. Erbert, „Generation of spectrally stable continuous-wave emission and ns pulses with a peak power of 4 W using a distributed Bragg reflector laser and a ridge-waveguide power amplifier“, *Optics Express*, vol. 22, no. 20, pp. 23 980–23 989, 2014.

[85] W.-C. Fang, C. Bethea, Y. Chen, and S. L. Chuang, „Longitudinal spatial inhomogeneities in high-power semiconductor lasers“, *IEEE Journal of Selected Topics in Quantum Electronics*, vol. 1, no. 2, pp. 117–128, 1995.

[86] J.-N. Fehr, M.-A. Dupertuis, T. P. Hessler, *et al.*, „Direct observation of longitudinal spatial hole burning in semiconductor optical amplifiers with injection“, *Applied Physics Letters*, vol. 78, no. 26, pp. 4079–4081, 2001.

[87] R. Pagano, M. Ziegler, J. W. Tomm, *et al.*, „Two-dimensional carrier density distribution inside a high power tapered laser diode", *Applied Physics Letters*, vol. 98, no. 22, p. 221 110, 2011.

[88] P. Albrodt, M. Niemeyer, M. Elattar, *et al.*, „Low-index quantum-barrier single-pass tapered semiconductor optical amplifiers for efficient coherent beam combining", *Semiconductor Science and Technology*, vol. 35, no. 6, p. 065 018, 2020.

[89] M. Radziunas, J. Fuhrmann, A. Zeghuzi, *et al.*, „Efficient coupling of dynamic electro-optical and heat-transport models for high-power broad-area semiconductor lasers", *Optical and Quantum Electronics*, vol. 51, no. 3, p. 69, 2019.

[90] G. P. Agrawal, „Fast-Fourier-transform based beam-propagation model for stripe-geometry semiconductor lasers: Inclusion of axial effects", *Journal of Applied Physics*, vol. 56, no. 11, pp. 3100–3109, 1984.

[91] R. Lang, A. Hardy, R. Parke, *et al.*, „Numerical analysis of flared semiconductor laser amplifiers", *IEEE Journal of Quantum Electronics*, vol. 29, no. 6, pp. 2044–2051, 1993.

[92] J. J. Lim, S. Sujecki, L. Lang, *et al.*, „Design and simulation of next-generation high-power, high-brightness laser diodes", *IEEE Journal of Selected Topics in Quantum Electronics*, vol. 15, no. 3, pp. 993–1008, 2009.

[93] M. Radziunas and R. Čiegis, „Effective numerical algorithm for simulations of beam stabilization in broad area semiconductor lasers and amplifiers", *Mathematical Modelling and Analysis*, vol. 19, no. 5, pp. 627–646, 2014.

[94] A. Zeghuzi, H. Wenzel, H.-J. Wünsche, M. Radziunas, U. Bandelow, and A. Knigge, „Modeling of current spreading in high-power broad-area lasers and its impact on the lateral far field divergence", in *Physics and Simulation of Optoelectronic Devices XXVI*, vol. 10526, SPIE, 2018, p. 105261H.

[95] A. Zeghuzi, M. Radziunas, H.-J. Wünsche, *et al.*, „Traveling wave analysis of non-thermal far-field blooming in high-power broad-area lasers", *IEEE Journal of Quantum Electronics*, vol. 55, no. 2, pp. 1–7, 2019.

[96] A. Zeghuzi, H.-J. Wünsche, H. Wenzel, *et al.*, „Time-dependent simulation of thermal lensing in high-power broad-area semiconductor lasers", *IEEE Journal of Selected Topics in Quantum Electronics*, vol. 25, no. 6, pp. 1–10, Nov. 2019.

[97] A. Zeghuzi, „Analysis of spatio-temporal phenomena in high-brightness diode lasers using numerical simulations", Ph.D. dissertation, Humboldt-Universität zu Berlin, Berlin, 2020.

[98] J. Buus, „Beamwidth for asymmetric and multilayer semiconductor laser structures", *IEEE Journal of Quantum Electronics*, vol. 17, no. 5, pp. 732–736, 1981.

[99] W. B. Joyce, „Current-crowded carrier confinement in double-heterostructure lasers", *Journal of Applied Physics*, vol. 51, no. 5, p. 2394, 1980.

[100] K. Unger, „Spontane und induzierte Emission in Laserdioden", *Zeitschrift für Physik*, vol. 207, pp. 322–331, 1967.

[101] H. Wenzel, „Basic aspects of high-power semiconductor laser simulation“, *IEEE Journal of Selected Topics in Quantum Electronics*, vol. 19, no. 5, pp. 1–13, 2013.

[102] F. Bugge, R. Bege, G. Blume, *et al.*, „Lifetime behavior of laser diodes with highly strained InGaAs QWs and emission wavelength between 1120 nm and 1180 nm“, *Journal of Crystal Growth*, vol. 491, pp. 31–35, 2018.

[103] N. Werner, J. Wegemund, S. Gerke, *et al.*, „Comparison of distributed Bragg reflector ridge waveguide diode lasers and monolithic master oscillator power amplifiers“, in *Novel In-Plane Semiconductor Lasers XVII*, vol. 10553, SPIE, 2018, p. 105531D.

[104] M. Kihara, S. Nagasawa, and T. Tanifuji, „Return loss characteristics of optical fiber connectors“, *Journal of Lightwave Technology*, vol. 14, no. 9, pp. 1986–1991, 1996.

[105] P. Chanclou, M. Thual, and J. Iostec, „High return loss at the end face of fiber“, *Applied Optics*, vol. 40, no. 4, pp. 458–460, 2001.

[106] D. J. Bossert, J. R. Marciante, and M. W. Wright, „Feedback effects in tapered broad-area semiconductor lasers and amplifiers“, *IEEE Photonics Technology Letters*, vol. 7, no. 5, pp. 470–472, 1995.

[107] S. Delepine, F. Gerard, A. Pinquier, *et al.*, „How to launch 1 W into single-mode fiber from a single 1.48-µm flared resonator“, *IEEE Journal of Selected Topics in Quantum Electronics*, vol. 7, no. 2, pp. 111–123, 2001.

[108] M. Hempel, M. Chi, P. M. Petersen, U. Zeimer, and J. W. Tomm, „How does external feedback cause AlGaAs-based diode lasers to degrade?“, *Applied Physics Letters*, vol. 102, no. 2, p. 023 502, 2013.

[109] B. Leonhäuser, H. Kissel, J. W. Tomm, M. Hempel, A. Unger, and J. Biesenbach, „High-power diode lasers under external optical feedback“, in *High-Power Diode Laser Technology and Applications XIII*, vol. 9348, SPIE, 2015, p. 93480M.

[110] M. A. Helal, S. N. Nyirenda-Kaunga, S. Bull, and E. C. Larkins, „Impact of unintentional external feedback on the performance of high-power tapered lasers“, in *2017 IEEE High Power Diode Lasers and Systems Conference (HPD)*, IEEE, 2017, pp. 21–22.

[111] S. Rauch, C. Holly, and H. Zimer, „Catastrophic optical damage in 950-nm broad-area laser diodes due to misaligned optical feedback and injection“, *IEEE Journal of Quantum Electronics*, vol. 54, no. 4, pp. 1–7, 2018.

[112] L. Borruel, H. Odriozola, J. M. G. Tijero, I. Esquivias, S. Sujecki, and E. C. Larkins, „Design strategies to increase the brightness of gain guided tapered lasers“, *Optical and Quantum Electronics*, vol. 40, no. 2, pp. 175–189, 2008.

[113] A. Siegman, „A canonical formulation for analyzing multielement unstable resonators“, *IEEE Journal of Quantum Electronics*, vol. 12, no. 1, pp. 35–40, 1976.

[114] N. Hodgson and H. Weber, *Laser Resonators and Beam Propagation: Fundamentals, Advanced Concepts and Applications*, 2nd ed. Springer New York, 2005.

[115] S. A. Collins, „Lens-system diffraction integral written in terms of matrix optics“, *Journal of the Optical Society of America*, vol. 60, no. 9, pp. 1168–1177, 1970.

[116] B. Eppich, „Die Charakterisierung von Strahlungsfeldern mit der Wigner-Verteilung und deren Messung“, Ph.D. dissertation, Technische Universität Berlin, Berlin, 1998.

[117] M. J. Bastiaans, „Wigner distribution function and its application to first-order optics“, *Journal of the Optical Society of America*, vol. 69, no. 12, p. 1710, 1979.

[118] M. A. Porras, „Nonspecular reflection of general light beams at a dielectric interface“, *Optics Communications*, vol. 135, no. 4, pp. 369–377, 1997.

[119] P. Hildenstein, N. Werner, K. Paschke, and G. Tränkle, „Investigation of astigmatism in tapered edge-emitting diode amplifiers at 980 nm“, in *Novel In-Plane Semiconductor Lasers XXI*, vol. 12021, SPIE, 2022, p. 120210G.

[120] R. Lang and K. Kobayashi, „External optical feedback effects on semiconductor injection laser properties“, *IEEE Journal of Quantum Electronics*, vol. 16, no. 3, pp. 347–355, 1980.

[121] T. Kurosaki, T. Hirono, and M. Fukuda, „Suppression of external cavity modes in DFB lasers with a high endurance against optical feedback“, *IEEE Photonics Technology Letters*, vol. 6, no. 8, pp. 900–902, 1994.

[122] M. Christensen, C. Zink, M. T. Jamal, A. K. Hansen, O. B. Jensen, and B. Sumpf, „Measuring the sensitivity to optical feedback of single-frequency high-power laser diodes“, *Journal of the Optical Society of America B*, vol. 38, no. 3, p. 885, 2021.

[123] R. Tkach and A. Chraplyvy, „Regimes of feedback effects in 1.5-µm distributed feedback lasers“, *Journal of Lightwave Technology*, vol. 4, no. 11, pp. 1655–1661, 1986.

[124] S. Donati and R.-H. Horng, „The diagram of feedback regimes revisited“, *IEEE Journal of Selected Topics in Quantum Electronics*, vol. 19, no. 4, p. 1 500 309, 2013.

[125] N. Schunk and K. Petermann, „Numerical analysis of the feedback regimes for a single-mode semiconductor laser with external feedback“, *IEEE Journal of Quantum Electronics*, vol. 24, no. 7, pp. 1242–1247, 1988.

[126] M. Norgia, G. Giuliani, and S. Donati, „Absolute distance measurement with improved accuracy using laser diode self-mixing interferometry in a closed loop“, *IEEE Transactions on Instrumentation and Measurement*, vol. 56, no. 5, pp. 1894–1900, 2007.

[127] D. Lenstra, B. Verbeek, and A. Den Boef, „Coherence collapse in single-mode semiconductor lasers due to optical feedback“, *IEEE Journal of Quantum Electronics*, vol. 21, no. 6, pp. 674–679, 1985.

[128] B. Kim, A. Locquet, D. Choi, and D. S. Citrin, „Experimental route to chaos of an external-cavity semiconductor laser“, *Physical Review A*, vol. 91, no. 6, p. 061 802, 2015.

[129] D. M. Kane and K. A. Shore, *Unlocking Dynamical Diversity: Optical Feedback Effects on Semiconductor Lasers.* John Wiley & Sons, Ltd, 2005.

[130] J. Houlihan, G. Huyet, and J. G. McInerney, „Dynamics of a semiconductor laser with incoherent optical feedback“, *Optics Communications*, vol. 199, no. 1, pp. 175–179, 2001.

[131] R. Ju and P. Spencer, „Dynamic regimes in semiconductor lasers subject to incoherent optical feedback“, *Journal of Lightwave Technology*, vol. 23, no. 8, pp. 2513–2523, 2005.

[132] Z. Zhang, H. Wang, N. Satyan, G. Rakuljic, C. T. Santis, and A. Yariv, „Coherent and incoherent optical feedback sensitivity of high-coherence Si/III-V hybrid lasers“, in *2019 Optical Fiber Communications Conference and Exhibition (OFC)*, Optica Publishing Group, 2019, p. W4E.3.

[133] B. Tykalewicz, D. Goulding, S. P. Hegarty, *et al.*, „Emergence of resonant mode-locking via delayed feedback in quantum dot semiconductor lasers“, *Optics Express*, vol. 24, no. 4, p. 4239, 2016.

[134] C. Schelte, P. Camelin, M. Marconi, *et al.*, „Third order dispersion in time-delayed systems“, *Physical Review Letters*, vol. 123, no. 4, p. 043 902, 2019.

[135] A. G. Vladimirov and D. Turaev, „Model for passive mode locking in semiconductor lasers“, *Physical Review A*, vol. 72, no. 3, p. 033 808, 2005.

[136] J. Mulet and S. Balle, „Mode-locking dynamics in electrically driven vertical-external-cavity surface-emitting lasers“, *IEEE Journal of Quantum Electronics*, vol. 41, no. 9, pp. 1148–1156, 2005.

[137] G. Acket, D. Lenstra, A. Den Boef, and B. Verbeek, „The influence of feedback intensity on longitudinal mode properties and optical noise in index-guided semiconductor lasers“, *IEEE Journal of Quantum Electronics*, vol. 20, no. 10, pp. 1163–1169, 1984.

[138] C. Zink, M. Christensen, M. T. Jamal, *et al.*, „Investigation of controlled external feedback on the properties of low and high-power frequency-stabilized diode laser“, in *Novel In-Plane Semiconductor Lasers XVIII*, vol. 10939, SPIE, 2019, p. 109391J.

[139] B. Tromborg, J. Osmundsen, and H. Olesen, „Stability analysis for a semiconductor laser in an external cavity“, *IEEE Journal of Quantum Electronics*, vol. 20, no. 9, pp. 1023–1032, 1984.

[140] C. Henry and R. Kazarinov, „Instability of semiconductor lasers due to optical feedback from distant reflectors“, *IEEE Journal of Quantum Electronics*, vol. 22, no. 2, pp. 294–301, 1986.

[141] V. Tronciu, N. Werner, H. Wenzel, and H.-J. Wünsche, „Feedback sensitivity of detuned DBR semiconductor lasers“, *IEEE Journal of Quantum Electronics*, vol. 57, no. 5, pp. 1–7, 2021.

[142] S. Bauer, O. Brox, J. Kreissl, *et al.*, „Nonlinear dynamics of semiconductor lasers with active optical feedback", *Physical Review E*, vol. 69, no. 1, p. 016 206, 2004.

[143] B. Tromborg and J. Mork, „Stability analysis and the route to chaos for laser diodes with optical feedback", *IEEE Photonics Technology Letters*, vol. 2, no. 8, pp. 549–552, 1990.

[144] T. Hirono, T. Kurosaki, and M. Fukuda, „Transition from the lowest linewidth mode operation to coherence collapse in a semiconductor laser with feedback from a distant reflector", *IEEE Journal of Quantum Electronics*, vol. 32, no. 5, pp. 829–834, 1996.

[145] N. Schunk and K. Petermann, „Stability analysis for laser diodes with short external cavities", *IEEE Photonics Technology Letters*, vol. 1, no. 3, pp. 49–51, 1989.

[146] R. J. Jones, P. S. Spencer, J. Lawrence, and D. M. Kane, „Influence of external cavity length on the coherence collapse regime in laser diodes subject to optical feedback", *IEE Proceedings - Optoelectronics*, vol. 148, no. 1, pp. 7–12, 2001.

[147] G. Morthier, „Feedback sensitivity of DBR-type laser diodes", *IEEE Photonics Journal*, vol. 13, no. 4, pp. 1–5, 2021.

[148] N. Umemura, D. Matsuda, T. Mizuno, and K. Kato, „Sellmeier and thermo-optic dispersion formulas for the extraordinary ray of 5 mol % MgO-doped congruent $LiNbO_3$ in the visible, infrared, and terahertz regions", *Applied Optics*, vol. 53, no. 25, pp. 5726–5732, 2014.

[149] N. Umemura, Y. Nakahara, J. Hirohashi, and Y. Furukawa, „Refined Sellmeier equation for the extraordinary wave in MgO doped stoichiometric $LiTaO_3$", in *Laser Congress 2020 (ASSL, LAC)*, Optica Publishing Group, 2020, p. JTh6A.24.

[150] M. Triplett, J. Khaydarov, X. Xu, *et al.*, „Multi-watt, broadband second-harmonic-generation in MgO:PPSLT waveguides fabricated with femtosecond laser micro-machining", *Optics Express*, vol. 27, no. 15, p. 21 102, 2019.

[151] A. W. Snyder and W. R. Young, „Modes of optical waveguides", *Journal of the Optical Society of America*, vol. 68, no. 3, pp. 297–309, 1978.

[152] A. G. Fox and T. Li, „Resonant modes in a maser interferometer", *The Bell System Technical Journal*, vol. 40, no. 2, pp. 453–488, 1961.

[153] N. Werner, P. Hildenstein, H. Wenzel, K. Paschke, and G. Tränkle, „Simulation of optical feedback propagating through tapered amplifiers", in *Physics and Simulation of Optoelectronic Devices XXX*, vol. 11995, SPIE, 2022, p. 1 199 503.

[154] Y. S. Kim and R. T. Smith, „Thermal expansion of lithium tantalate and lithium niobate single crystals", *Journal of Applied Physics*, vol. 40, no. 11, pp. 4637–4641, 1969.

[155] K. Liu and M. G. Littman, „Novel geometry for single-mode scanning of tunable lasers", *Optics Letters*, vol. 6, no. 3, pp. 117–118, 1981.

List of Publications

The research activities of this thesis have contributed to the following publications.

Journal Articles

- **N. Werner**, J. Hofmann, A. Sahm, *et al.*, „Direct modulation capabilities of micro-integrated laser sources in the yellow-green spectral range“, *IEEE Photonics Technology Letters*, vol. 29, no. 12, pp. 995–998, 2017
- K. Paschke, G. Blume, **N. Werner**, *et al.*, „Compact RGBY light sources with high luminance for laser display applications“, *Optical Review*, vol. 25, no. 1, pp. 149–159, 2017
- **N. Werner**, G. Blume, D. Feise, *et al.*, „Spectral mode hop characteristics of ridge waveguide lasers with distributed Bragg-reflector“, *IEEE Photonics Technology Letters*, vol. 29, no. 24, pp. 2183–2186, 2017
- A. Sahm, **N. Werner**, J. Hofmann, *et al.*, „Compact miniaturized laser module emitting more than 1.6 W of yellow light at 576 nm“, *IEEE Photonics Technology Letters*, vol. 30, no. 21, pp. 1878–1881, 2018
- **N. Werner**, J. Wegemund, D. Feise, *et al.*, „Emission behavior of distributed Bragg-reflector ridge waveguide lasers exposed to strong optical feedback“, *Applied Optics*, vol. 59, no. 28, pp. 8653–8660, 2020
- V. Tronciu, **N. Werner**, H. Wenzel, *et al.*, „Feedback sensitivity of detuned DBR semiconductor lasers“, *IEEE Journal of Quantum Electronics*, vol. 57, no. 5, pp. 1–7, 2021
- V. Tronciu, **N. Werner**, H. Wenzel, *et al.*, „Theoretical study of the behavior of a DBR laser subject to external optical feedback“, *Romanian Reports in Physics*, vol. 73, no. 408, pp. 1–12, 2021
- E. Grigoriev, V. Tronciu, **N. Werner**, *et al.*, „The influence of a residual reflectivity at the front facet of a multisection master-oscillator power-amplifier“, *Journal of Engineering Science*, vol. 29, no. 2, pp. 62–67, 2022
- **N. Werner**, N. Thieß, K. Paschke, *et al.*, „Optical feedback generated by tapered amplifiers emitting at 1120 nm“, *IEEE Journal of Quantum Electronics*, vol. 58, no. 2, pp. 1–8, 2022

- P. Hildenstein, D. Feise, F. Mauerhoff, **N. Werner**, *et al.*, „Switchable thermal waveguides in GaAs based devices“, *AIP Advances*, vol. 13, no. 2, p. 025 358, 2023

International Conference Contributions

- D. Jedrzejczyk, J. Hofmann, **N. Werner**, *et al.*, „Compact diode laser module at 1116 nm with an integrated optical isolation and a PM-SMF output“, in *Components and Packaging for Laser Systems III*, (San Francisco, California, United States, Feb. 22, 2017), vol. 10085, SPIE, 2017, p. 1 008 506

- J. Hofmann, D. Jedrzejczyk, A. Sahm, **N. Werner**, *et al.*, „Miniaturized laser amplifier modules for wavelengths of 1180 nm with PM-fiber input and more than 1 W optical output power“, in *High-Power Diode Laser Technology XV*, (San Francisco, California, United States, Feb. 22, 2017), vol. 10086, SPIE, 2017, p. 100860H

- **N. Werner**, G. Blume, D. Feise, *et al.*, „Master oscillator power amplifier concepts with nearly diffraction-limited watt-level continuous wave emission at 635 nm for laser projection“, presented at the 6th Laser Display and Lighting Conference (LDC'17) (Yokohama, Japan, Apr. 20, 2017)

- **N. Werner**, J. Wegemund, S. Gerke, *et al.*, „Comparison of distributed Bragg reflector ridge waveguide diode lasers and monolithic master oscillator power amplifiers“, in *Novel In-Plane Semiconductor Lasers XVII*, (San Francisco, California, United States, Feb. 19, 2018), vol. 10553, SPIE, 2018, p. 105531D

- **N. Werner**, A. Sahm, R. Bege, *et al.*, „Multi watt-level picosecond micro-laser sources in the yellow-green spectral range“, in *Nonlinear Frequency Generation and Conversion: Materials and Devices XVIII*, (San Francisco, California, United States, Mar. 4, 2019), vol. 10902, SPIE, 2019, p. 109021T

- A. Sahm, **N. Werner**, J. Hofmann, *et al.*, „Miniaturized watt-level laser modules emitting in the yellow-green spectral range for biophotonic applications“, in *Nonlinear Frequency Generation and Conversion: Materials and Devices XVIII*, (San Francisco, California, United States, Mar. 4, 2019), vol. 10902, SPIE, 2019, p. 109020C

- D. Feise, D. Jedrzejczyk, D. Krug, **N. Werner**, *et al.*, „High spectral radiance distributed Bragg reflector tapered diode lasers at 1060 nm with novel internal output DBR-grating“, in *Novel In-Plane Semiconductor Lasers XVIII*, (San Francisco, California, United States, Mar. 1, 2019), vol. 10939, SPIE, 2019, p. 109391F

- **N. Werner**, S. Häuser, and K. Paschke, „Experimental investigation of optical feedback from periodically poled crystals for nonlinear frequency conversion“, presented at the 2021 Conference on Lasers and Electro-Optics Europe / European Quantum Electronics Conference (CLEO/Europe-EQEC) (online, Jun. 24, 2021)

- P. Hildenstein, **N. Werner**, K. Paschke, *et al.*, „Investigation of astigmatism in tapered edge-emitting diode amplifiers at 980 nm", in *Novel In-Plane Semiconductor Lasers XXI,* (online, Mar. 3, 2022), vol. 12021, SPIE, 2022, p. 120210G
- **N. Werner**, P. Hildenstein, H. Wenzel, *et al.*, „Simulation of optical feedback propagating through tapered amplifiers", in *Physics and Simulation of Optoelectronic Devices XXX,* (online, Mar. 4, 2022), vol. 11995, SPIE, 2022, p. 1 199 503
- **N. Werner**, M. Triebkorn, A. Ginolas, *et al.*, „Efficiency optimization of ridge waveguide amplifiers at 1122 nm", in *Novel In-Plane Semiconductor Lasers XXII,* (San Francisco, California, United States, Feb. 2, 2023), vol. 12440, SPIE, 2023, p. 124 400

Patent Application

- **N. Werner**, J. Hofmann, R. Bege, *et al.*, „Device for generating laser radiation", WO2020104519, Nov. 23, 2018

Danksagung

Die Forschungsarbeiten, die dieser Arbeit zugrunde liegen, wurden am Ferdinand-Braun-Institut, Leibniz-Institut für Höchstfrequenztechnik in der Gruppe Laser Modules Lab durchgeführt. Sowohl am Institut als auch darüberhinausgehend konnte ich mich auf vielfältige Unterstützung verlassen. Ohne diese wäre die Arbeit in dieser Form nicht möglich gewesen. Dafür möchte ich mich ganz herzlich bedanken.

An erster Stelle gilt mein Dank Prof. Dr. Günther Tränkle für die Möglichkeit am Ferdinand-Braun-Institut forschen zu können und für die Betreuung und Begutachtung dieser Arbeit. Die zahlreichen fachlichen Diskussionen waren stets hilfreich und haben neue wissenschaftliche Impulse gegeben.

Bedanken möchte ich mich auch bei Prof. Dr. Carlo Holly für die Bereitschaft zur Begutachtung dieser Arbeit. In gleicher Weise danke ich Prof. Achim Peters, Ph.D. dafür, dass er sich bereiterklärt hat, die Arbeit zu begutachten.

Bei Dr. Katrin Paschke möchte ich mich für die Betreuung der Arbeit bedanken. Die ständige Hilfsbereitschaft bei fachlichen und organisatorischen Fragen sowie der offene Umgang miteinander im Laser Modules Lab waren eine große Stütze. Besonderer Dank für die Betreuung gilt auch Dr. Hans Wenzel. Die vielen Anregungen und Diskussionen waren von großem Wert insbesondere bei der theoretischen Modellierung.

Den Kollegen des Laser Modules Labs danke ich für den Zusammenhalt und die unzähligen fachlichen und nichtfachlichen Gespräche. Hervorzuheben sind hier Dr. Gunnar Blume, Dr. David Feise, Alexander Sahm, Philipp Hildenstein und Felix Mauerhoff. Ferner möchte ich Jan Wegemund, Nina Thieß und Selina Häuser nicht unerwähnt lassen, die mit ihren Abschlussarbeiten und die Arbeit als studentische Hilfskraft eine große Unterstützung im Labor waren.

Ich möchte mich auch bei allen anderen Kolleginnen und Kollegen des Ferdinand-Braun-Institutes bedanken, die direkt oder indirekt zum Erfolg dieser Arbeit beigetragen haben. Besonderer Dank geht an Dr. Bernd Eppich für die vielen aufschlussreichen Diskussionen. Darüber hinaus möchte ich Dr. Jan-Philipp Koester und Alina Hahn dankend erwähnen.

Meiner Familie danke ich für das Vertrauen und dass sie mich von Anfang an auf meinem Weg unterstützt und bestärkt hat. Dankbar bin ich auch meinen Freunden, die immer für mich da sind, auch über Landesgrenzen hinweg. Abschließend gilt besonderer Dank meiner Partnerin Sophie für die vielen schönen Momente, die mir Kraft geben.

Berlin, 5. Juli 2023
Nils Werner

Innovationen mit Mikrowellen und Licht
Forschungsberichte aus dem Ferdinand-Braun-Institut, Leibniz-Institut für Höchstfrequenztechnik

Herausgeber: Prof. Dr.-Ing Patrick Scheele

Band 1: **Thorsten Tischler**
Die Perfectly-Matched-Layer-Randbedingung in der Finite-Differenzen-Methode im Frequenzbereich: Implementierung und Einsatzbereiche
ISBN: 3-86537-113-2, 19,00 EUR, 144 Seiten

Band 2: **Friedrich Lenk**
Monolithische GaAs FET- und HBT-Oszillatoren mit verbesserter Transistormodellierung
ISBN: 3-86537-107-8, 19,00 EUR, 140 Seiten

Band 3: **R. Doerner, M. Rudolph (eds.)**
Selected Topics on Microwave Measurements, Noise in Devices and Circuits, and Transistor Modeling
ISBN: 3-86537-328-3, 19,00 EUR, 130 Seiten

Band 4: **Matthias Schott**
Methoden zur Phasenrauschverbesserung von monolithischen Millimeterwellen-Oszillatoren
ISBN: 978-3-86727-774-0, 19,00 EUR, 134 Seiten

Band 5: **Katrin Paschke**
Hochleistungsdiodenlaser hoher spektraler Strahldichte mit geneigtem Bragg-Gitter als Modenfilter (α-DFB-Laser)
ISBN: 978-3-86727-775-7, 19,00 EUR, 128 Seiten

Band 6: **Andre Maaßdorf**
Entwicklung von GaAs-basierten Heterostruktur-Bipolartransistoren (HBTs) für Mikrowellenleistungszellen
ISBN: 978-3-86727-743-3, 23,00 EUR, 154 Seiten

Band 7: **Prodyut Kumar Talukder**
Finite-Difference-Frequency-Domain Simulation of Electrically Large Microwave Structures using PML and Internal Ports
ISBN: 978-3-86955-067-1, 19,00 EUR, 138 Seiten

Band 8: **Ibrahim Khalil**
Intermodulation Distortion in GaN HEMT
ISBN: 978-3-86955-188-3, 23,00 EUR, 158 Seiten

Band 9: **Martin Maiwald**
Halbleiterlaser basierte Mikrosystemlichtquellen für die Raman-Spektroskopie
ISBN: 978-3-86955-184-5, 19,00 EUR, 134 Seiten

Band 10: **Jens Flucke**
Mikrowellen-Schaltverstärker in GaN- und GaAs-Technologie Designgrundlagen und Komponenten
ISBN: 978-3-86955-304-7, 21,00 EUR, 122 Seiten

Cuvillier Verlag
Internationaler wissenschaftlicher Fachverlag

Innovationen mit Mikrowellen und Licht
Forschungsberichte aus dem Ferdinand-Braun-Institut, Leibniz-Institut für Höchstfrequenztechnik

Herausgeber: Prof. Dr.-Ing Patrick Scheele

Band 11: **Harald Klockenhoff**
Optimiertes Design von Mikrowellen-Leistungstransistoren und Verstärkern im X-Band
ISBN: 978-3-86955-391-7, 26,75 EUR, 130 Seiten

Band 12: **Reza Pazirandeh**
Monolithische GaAs FET- und HBT-Oszillatoren mit verbesserter Transistormodellierung
ISBN: 978-3-86955-107-8, 19,00 EUR, 140 Seiten

Band 13: **Tomas Krämer**
High-Speed InP Heterojunction Bipolar Transistors and Integrated Circuits in Transferred Substrate Technology
ISBN: 978-3-86955-393-1, 21,70 EUR, 140 Seiten

Band 14: **Phuong Thanh Nguyen**
Investigation of spectral characteristics of solitary diode lasers with integrated grating resonator
ISBN: 978-3-86955-651-2, 24,00 EUR, 156 Seiten

Band 15: **Sina Riecke**
Flexible Generation of Picosecond Laser Pulses in the Infrared and Green Spectral Range by Gain-Switching of Semiconductor Lasers
ISBN: 978-3-86955-652-9, 22,60 EUR, 136 Seiten

Band 16: **Christian Hennig**
Hydrid-Gasphasenepitaxie von versetzungsarmen und freistehenden GaN-Schichten
ISBN: 978-3-86955-822-6, 27,00 EUR, 162 Seiten

Band 17: **Tim Wernicke**
Wachstum von nicht- und semipolaren InAlGaN-Heterostrukturen für hocheffiziente Licht-Emitter
ISBN: 978-3-86955-881-3, 23,40 EUR, 138 Seiten

Band 18: **Andreas Wentzel**
Klasse-S Mikrowellen-Leistungsverstärker mit GaN-Transistoren
ISBN: 978-3-86955-897-4, 29,65 EUR, 172 Seiten

Band 19: **Veit Hoffmann**
MOVPE growth and characterization of (In,Ga)N quantum structures for laser diodes emitting at 440 nm
ISBN: 978-3-86955-989-6, 18,00 EUR, 118 Seiten

Band 20: **Ahmad Ibrahim Bawamia**
Improvement of the beam quality of high-power broad area semiconductor diode lasers by means of an external resonator
ISBN: 978-3-95404-065-0, 21,00 EUR, 126 Seiten

Cuvillier Verlag
Internationaler wissenschaftlicher Fachverlag

Innovationen mit Mikrowellen und Licht
Forschungsberichte aus dem Ferdinand-Braun-Institut, Leibniz-Institut für Höchstfrequenztechnik

Herausgeber: Prof. Dr.-Ing Patrick Scheele

Band 21: **Agnietzka Pietrzak**
Realization of High Power Diode Lasers with Extremely Narrow Vertical Divergence
ISBN: 978-3-95404-066-7, 27,40 EUR, 144 Seiten

Band 22: **Eldad Bahat-Treidel**
GaN-based HEMTs for High Voltage Operation
Design, Technology and Characterization
ISBN: 978-3-95404-094-0, 41,10 EUR, 220 Seiten

Band 23: **Ponky Ivo**
AlGaN/GaN HEMTs Reliability:
Degradation Modes and Anslysis
ISBN: 978-3-95404-259-3, 23,55 EUR, 132 Seiten

Band 24: **Stefan Spießberger**
Compact Semiconductor-Based Laser Sources
with Narrow Linewidth and High Output Power
ISBN: 978-3-95404-261-6, 24,15 EUR, 140 Seiten

Band 25: **Silvio Kühn**
Mikrowellenoszillatoren für die Erzeugung von atmosphärischen Mikroplasmen
ISBN: 978-3-95404-378-1, 21,85 EUR, 112 Seiten

Band 26: **Sven Schwertfeger**
Experimentelle Untersuchung der Modensynchronisation in Multisegment-Laserdioden zur Erzeugung kurzer optischer Pulse bei einer Wellenlänge von 920 nm
ISBN: 978-3-95404-471-9, 29,45 EUR, 150 Seiten

Band 27: **Christoph Matthias Schultz**
Analysis and mitigation of the factors limiting the effiency of high power distributed feedback diode lasers
ISBN: 978-3-95404-521-1, 68,40 EUR, 388 Seiten

Band 28: **Luca Redaelli**
Design and fabrication of GaN-based laser diodes for single-mode and narrow-linewidth applications
ISBN: 978-3-95404-586-0, 29,70 EUR, 176 Seiten

Band 29: **Martin Spreemann**
Resonatorkonzepte für Hochleistungs-Diodenlaser
mit ausgedehnten lateralen Dimensionen
ISBN: 978-3-95404-628-7, 25,15 EUR, 128 Seiten

Cuvillier Verlag
Internationaler wissenschaftlicher Fachverlag

Innovationen mit Mikrowellen und Licht
Forschungsberichte aus dem Ferdinand-Braun-Institut, Leibniz-Institut für Höchstfrequenztechnik

Herausgeber: Prof. Dr.-Ing Patrick Scheele

Band 30: **Christian Fiebig**
Diodenlaser mit Trapezstruktur und hoher Brillanz für die Realisierung einer Frequenzkonversion auf einer mikro-optischen Bank
ISBN: 978-3-95404-690-4, 26,30 EUR, 140 Seiten

Band 31: **Viola Küller**
Versetzungsreduzierte AIN- und AlGaN-Schichten als Basis für UV LEDs
ISBN: 978-3-95404-741-3, 34,40 EUR, 164 Seiten

Band 32: **Daniel Jedrzejczyk**
Efficient frequency doubling of near-infrared diode lasers using quasi phase-matched waveguides
ISBN: 978-3-95404-958-5, 27,90 EUR, 134 Seiten

Band 33: **Sylvia Hagedorn**
Hybrid-Gasphasenepitaxie zur Herstellung von Aluminiumgalliumnitrid
ISBN: 978-3-95404-985-1, 38,00 EUR, 176 Seiten

Band 34: **Alexander Kravets**
Advanced Silicon MMICs for mm-Wave Automotive Radar Front-Ends
ISBN: 978-3-95404-986-8, 31,90 EUR, 156 Seiten

Band 35: **David Feise**
Longitudinale Modenfilter für Kantanemitter im roten Spektralbereich
ISBN: 978-3-7369-9116-3, 39,20 EUR, 168 Seiten

Band 36: **Ksenia Nosaeva**
Indium phosphide HBT in thermally optimized periphery for applications up to 300GHZ
ISBN: 978-3-7369-287-0, 42,00 EUR, 154 Seiten

Band 37: **Muhammad Maruf Hossain**
Signal Generation for Millimeter Wave and THZ Applications in InP-DHBT and InP-on-BiCMOS Technologies
ISBN: 978-3-7369-9335-8, 35,60 EUR, 136 Seiten

Band 38: **Sirinpa Monayakul**
Development of Sub-mm Wave Flip-Chip Interconnect
ISBN: 978-3-7369-9410-2, 44,00 EUR, 146 Seiten

Band 39: **Moritz Brendel**
Charakterisierung und Optimierung von (Al, Ga) N-basierten UV-Photodetektoren
ISBN: 978-3-7369-9465-2, 49,90 EUR, 196 Seiten

Band 40: **Erdenetsetseg Luvsandamdin**
Development of micro-integrated diode lasers for precision quantum optics experiments in space
ISBN: 978-3-7369-9479-9, 39,00 EUR, 126 Seiten

Cuvillier Verlag
Internationaler wissenschaftlicher Fachverlag

Innovationen mit Mikrowellen und Licht
Forschungsberichte aus dem Ferdinand-Braun-Institut, Leibniz-Institut für Höchstfrequenztechnik

Herausgeber: Prof. Dr.-Ing Patrick Scheele

Band 41: **Thi Nghiem Vu**
Development and analysis of diode laser ns-MOPA systems for high peak power application
ISBN: 978-3-7369-9480-5, 38,80 EUR, 138 Seiten

Band 42: **Christian Bansleben**
Differentieller Mikrowellen-Leistungsoszillator für die Realisierung ultrakompakter Plasmaquellen in Matrixanordnung
ISBN: 978-3-7369-9530-7, 34,90 EUR, 136 Seiten

Band 43: **Martin Winterfeldt**
Investigation of slow-axis beam quality degradation in high-power broad area diode lasers
ISBN: 978-3-7369-9733-2, 39,90 EUR, 158 Seiten

Band 44: **Jonathan Decker**
Investigation of monolithically integrated spectral stabilization in high-brightness broad area diode lasers
ISBN: 978-3-7369-9798-1, 49,50 EUR, 174 Seiten

Band 45: **Andreea Cristina Andrei**
Untersuchung und Optimierung robuster und hochlinearer rauscharmer Verstärker in GaN-Technologie
ISBN: 978-3-7369-9810-0, 39,90 EUR, 148 Seiten

Band 46: **Peng Luo**
GaN HEMT Modeling Including Trapping Effects Based on Chalmers Model and Pulsed S-Parameter Measurements
ISBN: 978-3-7369-9906-0, 48,00 EUR, 160 Seiten

Band 47: **Jörg Jeschke**
Entwicklung von optisch pumpbaren UVC-Lasern auf AlGaN-Basis
Chalmers Model and Pulsed S-Parameter Measurements
ISBN: 978-3-7369-9918-3, 44,90 EUR, 176 Seiten

Band 48: **Nikolai Wolff**
Wideband GaN Microwave Power Amplifiers with Class-G Supply Modulation
ISBN: 978-3-7369-9931-2, 44,90 EUR, 170 Seiten

Band 49: **Carlo Frevert**
Optimization of broad-area GaAs diode lasers for high powers and high efficiences in the temperature range 200-220 K
ISBN: 978-3-7369-9944-2, 44,90 EUR, 174 Seiten

Band 50: **Bassem Arar**
GaAs-based components for photonic integrated circuits
ISBN: 978-3-7369-9976-3, 43,60 EUR, 152 Seiten

Cuvillier Verlag
Internationaler wissenschaftlicher Fachverlag

Innovationen mit Mikrowellen und Licht
Forschungsberichte aus dem Ferdinand-Braun-Institut, Leibniz-Institut für Höchstfrequenztechnik

Herausgeber: Prof. Dr.-Ing Patrick Scheele

Band 51: **Mahmoud Tawfieq**
Development and characterisation of a diode laser based tunable high-power MOPA system
ISBN: 978-3-7369-9983-1, 44,90 EUR, 170 Seiten

Band 52: **Sebastian Preis**
Hocheffiziente frequenzagile Mikrowellen-Leistungsverstärker auf Basis von Verbindungshalbleitern und Ferroelektrika
ISBN: 978-3-7369-7004-5, 54,00 EUR, 136 Seiten

Band 53: **Simon Fleischmann**
Materialaspekte der Hydridgasphasenepitaxie von Aluminiumgalliumnitrid
ISBN: 978-3-7369-7029-8, 41,80 EUR, 160 Seiten

Band 54: **Erhan Ersoy**
Optimierung von koplanaren GaN-MMIC-Leistungsverstärkern im X-Band
ISBN: 978-3-7369-7157-8, 39,90 EUR, 154 Seiten

Band 55: **Simon Rauch**
Lateral emission characteristics of high-power broad-area lasers subject to external optical feedback
ISBN: 978-3-7369-7168-5, 29,90 EUR, 112 Seiten

Band 56: **Frank Dittmar**
Untersuchung der Strahlgüte von brillanten Hochleistungs-Trapezlasern für den Wellenlängenbereich bei 808 nm
ISBN: 978-3-7369-7167-7, 44,90 EUR, 176 Seiten

Band 57: **Florian Hühn**
Flexibler Modulator und Digitalverstärker-MMIC für den energieeffizienten Betrieb einer digitalen Sendekette im GHz-Bereich
ISBN: 978-3-7369-7190-5, 29,90 EUR, 162 Seiten

Band 58: **Dimitri Stoppel**
Interconnection development for InP-HBT terahertz circuits
ISBN: 978-3-7369-7204-9, 44,90 EUR, 156 Seiten

Band 59: **Anissa Zeghuzi**
Analysis of Spatio-Temporal Phenomena in High-Brightness Diode Lasers using Numerical Simulations
ISBN: 978-3-7369-7289-6, 49,90 EUR, 176 Seiten

Band 60: **Tasmim Alam**
Spectroscopic Applications of Terahertz Quantum-Cascade Lasers
ISBN: 978-3-7369-7297-1, 39,90 EUR, 132 Seiten

Innovationen mit Mikrowellen und Licht
Forschungsberichte aus dem Ferdinand-Braun-Institut, Leibniz-Institut für Höchstfrequenztechnik

Herausgeber: Prof. Dr.-Ing Patrick Scheele

Band 61: **Max Schiemangk**
Ein Lasersystem für Experimente mit Quantengasen unter Schwerelosigkeit
ISBN: 978-3-7369-7293-3, 49,90 EUR, 166 Seiten

Band 62: **Thorben Kaul**
Epitaxial Design Optimizations for Increased Efficiency in GaAs-Based High Power Diode Lasers
ISBN: 978-3-7369-7396-1, 38,90 EUR, 136 Seiten

Band 63: **Pietro della Casa**
Two Step MOPVE, in-situ etching and buried implantations: applications to the realization of GaAs laser diodes
ISBN: 978-3-7369-7397-8, 72,90 EUR, 250 Seiten

Band 64: **Heike Christopher**
A compact mode-locked diode laser system for high prescision frequency comparison experiments
ISBN: 978-3-7369-7399-2, 59,90 EUR, 206 Seiten

Band 65: **Rimma Zhytnytska**
Design von GaN Transistoren für leistungselektronische Anwendungen
ISBN: 978-3-7369-7413-5, 57,90 EUR, 182 Seiten

Band 66: **Sebastian Walde**
AlN base layers for UV LEDs
ISBN: 978-3-7369-7451-7, 44,90 EUR, 156 Seiten

Band 67: **Jan Ruschel**
Ursachen der stromgetriebenen Degradation von UV-LEDs
ISBN: 978-3-7369-7542-2, 49,90 EUR, 180 Seiten

Band 68: **Erik Freier**
Optimierung der Prozesstechnologie und Steigerung der Zuverlässungkeit und Lebensdauer von (InAlGa)N-basierten Halbleiterlaserdioden
ISBN: 978-3-7369-7604-7, 49,90 EUR, 178 Seiten

Band 69: **Norman Ruhnke**
A deep ultraviolet laser light source by frequency doubling of GaN based external cavity diode laser radiation
ISBN: 978-3-7369-7613-9, 39,90 EUR, 144 Seiten

Band 70: **Matthias M. Karow**
Broad-Area Laser Bars for 1 kW-Emission. Demonstrating increased efficiency and narrow far field
ISBN: 978-3-7369-7626-9, 39,50 EUR, 142 Seiten

Cuvillier Verlag
Internationaler wissenschaftlicher Fachverlag

Innovationen mit Mikrowellen und Licht
Forschungsberichte aus dem Ferdinand-Braun-Institut, Leibniz-Institut für Höchstfrequenztechnik

Herausgeber: Prof. Dr.-Ing Patrick Scheele

Band 71:	**Yi-Fan Tsao** Transceiver Technologies for Millimeter-Wave Beam Steering Applications ISBN: 978-3-7369-7702-0, 39,90 EUR, 146 Seiten
Band 72:	**Adam Rämer** Design und Realisierung GaN-HEMT basierter THz-Detektoren ISBN: 978-3-7369-7743-3, 58,90 EUR, 172 Seiten
Band 73:	**Jan-Philipp Koester** Design, simulation and analysis of laterally-longitudinally non-uniform edge-emitting GaAs-based diode lasers ISBN: 978-3-7369-7882-9 , 170 Seiten,
Band 74:	**Konstantin Osipov** Short channel GaN FET MMIC technology for high reliability applications ISBN: 978-3-7369-7956-7, 61,00 EUR, 170 Seiten
Band 75:	**Martin Kastner** Entwicklung aktiv gütegeschalteter Halbleiterlaser für LiDAR-Systeme des autonomen Fahrzeugs ISBN: 978-3-7369-7976-5, 40,00 EUR, 142 Seiten

Cuvillier Verlag
Internationaler wissenschaftlicher Fachverlag